# Gender and the Social Dimensio Climate Change

I0054554

Dispelling the myth that people in the Global North share similar experiences of climate change, this book reveals how intersecting social dimensions of climate change—people, processes, and institutions—give rise to different experiences of loss, adaptation, and resilience among those living in rural and resource contexts of the Global North.

Bringing together leading feminist researchers and practitioners from three countries—Australia, Canada, and Spain—this collection documents gender relations in fossil fuel, mining, and extractive industries, in land-based livelihoods, in approaches for inclusive environmental policy, and in the lived experience of climate hazards. Uniquely, the book brings together the voices, expertise, and experiences of both academic researchers and women whose views have *not* been prioritized in formal policies—for example, women in agriculture, Indigenous women, immigrant women, and women in male-dominated professions. Their contributions are insightful and compelling, highlighting the significance of gaining diverse perspectives for a fuller understanding of climate change impacts, more equitable processes and strategies for climate change adaptation, and a more welcoming climate future.

This book will be vital reading for students and scholars of gender studies, environmental studies, environmental sociology, geography, and sustainability science. It will provide important insights for planners, decision makers, and community advocates to strengthen their understanding of social dimensions of climate change and to develop more inclusive and equitable adaptation policies, plans, and practices.

**Amber J. Fletcher** is Associate Professor of Sociology & Social Studies and Academic Director of the Community Engagement and Research Centre at the University of Regina, Canada. Her research examines how gender and social inequality affect the lived experience of climate disaster in rural and Indigenous communities of the Canadian Prairies. In 2017, she edited the book *Women in Agriculture Worldwide* with Dr Wendee Kubik. She has served as a consultant to the United Nations World Water Assessment Programme and as an official delegate to the United Nations Commission on the Status of Women. She is a contributing author to a 2019 special report of the Intergovernmental Panel on Climate Change (IPCC) and is former President of the Canadian Research Institute for the Advancement of Women. She holds two medals from the Governor General of Canada for her research and advocacy on gender equality in Canada. In 2020 she was the Greeley Scholar for Peace Studies at the University of Massachusetts Lowell.

**Maureen G. Reed** is Distinguished Professor and a UNESCO Chair in Biocultural Diversity, Sustainability, Reconciliation, and Renewal at the University of Saskatchewan, Canada. Her research programme focuses on the social dimensions of sustainability—how people, processes, and institutions shape decisions about environment and development. She has received many awards for her work, including the Canadian Association of Geographers Award for Scholarly Distinction in Geography; the University of Saskatchewan Distinguished Researcher Award; the YWCA-Saskatoon Women of Distinction Lifetime Achievement Award and the University of Saskatchewan Distinguished Graduate Supervisor Award.

# Routledge Studies in Gender and Environments

With the European Union, United Nations, UN Framework Convention on Climate Change, and national governments and businesses at least ostensibly paying more attention to gender, including as it relates to environments, there is more need than ever for existing and future scholars, policy makers, and environmental professionals to understand and be able to apply these concepts to work towards greater gender equality in and for a sustainable world.

Comprising edited collections, monographs and textbooks, this new *Routledge Studies in Gender and Environments* series will incorporate sophisticated critiques and theorizations, including engaging with the full range of masculinities and femininities, intersectionality, and LBGTIQ perspectives. The concept of "environment" will also be drawn broadly to recognize how built, social, and natural environments intersect with and influence each other. Contributions will also be sought from global regions and contexts which are not yet well represented in gender and environments literature, in particular Russia, the Middle East, and China, as well as other East Asian countries such as Japan and Korea.

**Series Editor:** *Professor* **Susan Buckingham**, *an independent researcher, consultant, and writer on gender and environment related issues.*

**International Editorial Board**

**Margaret Alston** *is Professor of Social Work and Head of Department at Monash University, Melbourne, Australia.*
**Giovanna Di Chiro** *is Professor of Environmental Studies and teaches in the Gender and Sexuality Studies Program at Swarthmore College, USA.*
**Marjorie Griffin Cohen** *is an economist who is Professor Emeritus of Political Science and Gender, Sexuality and Women's Studies at Simon Fraser University, Canada.*
**Martin Hultman** *is a Senior Lecturer at Linköping University, Sweden.*
**Virginie Le Masson** *is a Research Fellow at the Overseas Development Institute, London, UK.*
**Sherilyn MacGregor** *is a Reader in Environmental Politics at the University of Manchester, UK.*
**Tanja Mölders** *is an environmental scientist. Since 2013 she is University Professor and holds the chair "Space and Gender" at Leibniz University, Hannover, Germany.*
**Karen Morrow** *is Professor of Environmental Law at Swansea University, UK.*
**Marion Roberts** *is Professor of Urban Design at Westminster University, UK.*

Titles in this series include:

For more information about this series, please visit: www.routledge.com/Routledge -Studies-in-Gender-and-Environments/book-series/RSGE

# Gender and the Social Dimensions of Climate Change

Rural and Resource Contexts of the Global North

**Edited by Amber J. Fletcher and Maureen G. Reed**

Routledge
Taylor & Francis Group
LONDON AND NEW YORK

earthscan
from Routledge

First published 2023
by Routledge
4 Park Square, Milton Park, Abingdon, Oxon OX14 4RN

and by Routledge
605 Third Avenue, New York, NY 10158

*Routledge is an imprint of the Taylor & Francis Group, an informa business*

© 2023 selection and editorial matter, Amber J. Fletcher and Maureen G. Reed; individual chapters, the contributors

The right of Amber J. Fletcher and Maureen G. Reed to be identified as the authors of the editorial material, and of the authors for their individual chapters, has been asserted in accordance with sections 77 and 78 of the Copyright, Designs and Patents Act 1988.

All rights reserved. No part of this book may be reprinted or reproduced or utilised in any form or by any electronic, mechanical, or other means, now known or hereafter invented, including photocopying and recording, or in any information storage or retrieval system, without permission in writing from the publishers.

*Trademark notice*: Product or corporate names may be trademarks or registered trademarks, and are used only for identification and explanation without intent to infringe.

*British Library Cataloguing-in-Publication Data*
A catalogue record for this book is available from the British Library

*Library of Congress Cataloging-in-Publication Data*
A catalog record has been requested for this book

ISBN: 978-0-367-54418-8 (hbk)
ISBN: 978-1-032-31685-7 (pbk)
ISBN: 978-1-003-08920-9 (ebk)

DOI: 10.4324/9781003089209

Typeset in Times New Roman
by Deanta Global Publishing Services, Chennai, India

# Contents

vi  *Contents*

# Figures

# Acknowledgements

This book was prepared during the COVID-19 pandemic, amidst a spate of record-setting wildfire and flood events in the countries discussed here. Despite the many challenges, the dedication of our contributing authors made this book a reality. We would like to sincerely thank all of the authors in this book for your important, thoughtful, and timely contributions regarding issues that become more pressing every day. We particularly thank those working on the front lines of climate hazards for making the time to share their insights with us.

We would also like to acknowledge the students and post-doctoral fellows who collectively form our research team on gender and climate change in Canada: Dr Nancy Sah Akwen, Dr John Boakye-Danquah, Dr Heidi Walker, Holly Campbell, Clodine Mbuli Shei, Angela Culham, Tina Elliott, Michaela Sidloski, and Nicholas Antonini. Thank you for your ongoing insights and dedication to the field.

Thank you to Marlis Merry for careful assistance with the formatting and references in this book.

We appreciate the work of our editors at Routledge—Annabelle Harris, Matthew Shobbrook, and Jyotsna Gurung—for their patience and guidance as we brought the book to realization. Thank you to the anonymous reviewers who took the time to provide valuable feedback.

We gratefully acknowledge funding from the Social Sciences and Humanities Research Council of Canada for supporting our research program on gender, intersectionality, and climate change.

Finally, we thank our families, including our four-legged members, for keeping us healthy and happy—especially during the past three years.

# Different Experiences, Diverse Knowledges

## Gender, Intersectionality, and Climate Change in Rural and Resource Contexts

*Amber J. Fletcher and Maureen G. Reed*

*Amber J. Fletcher is Associate Professor of Sociology & Social Studies and Academic Director of the Community Engagement and Research Centre at the University of Regina, Canada. Her research examines how gender and social inequality affect the lived experience of climate disaster in rural and Indigenous communities of the Canadian Prairies. Her research focuses, in particular, on the lives of farm women. In 2017, she edited the book* Women in Agriculture Worldwide *with Dr Wendee Kubik. She has served as a consultant to the United Nations World Water Assessment Programme and as an official delegate to the United Nations Commission on the Status of Women. She is a contributing author to a 2019 special report of the Intergovernmental Panel on Climate Change (IPCC) and is former President of the Canadian Research Institute for the Advancement of Women. She holds two medals from the Governor General of Canada for her research and advocacy on gender equality in Canada. In 2020 she was the Greeley Scholar for Peace Studies at the University of Massachusetts Lowell.*

*Maureen G. Reed is Distinguished Professor and a UNESCO Chair in Biocultural Diversity, Sustainability, Reconciliation, and Renewal at the University of Saskatchewan, Canada. Her research program focuses on the social dimensions of sustainability—how people, processes, and institutions shape decisions about environment and development. Research related to equity, diversity, and inclusion in Canada's forest sector focuses on how we can build processes for climate change adaptation and post-disaster learning that account for differences across gender and cultural diversity. She has received many awards for her work, including the Canadian Association of Geographers Award for Scholarly Distinction in Geography; the University of Saskatchewan Distinguished Researcher Award; the YWCA-Saskatoon*

DOI: 10.4324/9781003089209-1

*Women of Distinction Lifetime Achievement Award, and the University of Saskatchewan Distinguished Graduate Supervisor Award. When she is not learning from her research partners, colleagues, and students, she is walking the dog, taking photos of flowers, and getting her hands dirty in the garden.*

*Climate change affects us all.*

(former United Nations Secretary General
Ban Ki-moon, 2014)

Climate change *does* affect us all. Even in the Global North, we are experiencing unprecedented extreme events. In late 2019 and early 2020, just as the COVID-19 pandemic began to sweep the world, Australia was hit with one of its most severe wildfire seasons ever recorded (Deb et al., 2020). Caused in part by climate change-related drought, heat waves, and changes in vegetation (Deb et al., 2020), the "Black Summer" wildfires killed over three billion animals and 26 people, and also destroyed 5.5 million hectares of land and thousands of homes (Australian Disaster Resilience Knowledge Hub, n.d.) On the heels of the Australian wildfires, the year 2021 brought a spate of record-setting temperatures around the world, including the highest temperatures ever recorded in several American cities (Di Liberto, 2021), in Canada (Douty, 2021), and in the entire continent of Europe (Met Office, 2021). In the Canadian province of British Columbia, record summer temperatures were followed by record autumn rainfalls. Both resulted in massive losses, including damage to transportation corridors, other infrastructure, and agricultural lands. Over a million livestock and about 600 people perished. Direct economic losses have yet to be tallied, but are expected to run into billions of Canadian dollars.

But climate change *does not* affect us all equally. The burdens of climate change are not evenly distributed—either globally or within the Global North. People living in rural places, whose lives and livelihoods often rely directly on natural resources and the environment, are experiencing some of the most direct and dramatic effects of climate hazards. Many of these rural people face marginalization or exclusion due to their gender, racialized identity, socio-economic status, or political views. Climate hazards interact with the social dynamics of everyday life, often exacerbating such social inequality. Extreme situations simultaneously expose, intensify, and even challenge existing power differentials and inequalities *within* households and communities (Seager, 2014), as well as *between* households, communities, and governance authorities. It is not only the effects of climate change that are shaped by gender and other power dynamics, but also their causes and governance (MacGregor, 2010).

Nonetheless, the discourse of a shared experience continues to imply that climate change affects all women, men, and gender-diverse people equally (Bonewit

& Shreeves, 2015). This discourse then shapes actions. When formal governmental responses to climate change and climate hazards do not account for the inequalities and variations within communities, they merely add another layer to the burdens already experienced by marginalized groups in everyday life.

This book highlights the gender and social dimensions of climate change in three countries of the Global North. Our purpose is to reveal what can be learned through the application of a multi-level, intersectional lens. We do so in two ways. First, we draw attention to how people's experiences of climate change and climate hazards are shaped by intersecting elements of social identity (e.g., gender, race, rurality) and structural social relations of power (e.g., policy, law, ideology). Second, by including experiential and practitioner reflections on each of the contributed chapters, we bring the voices of women with first-hand knowledge of climate, gender, and environmental issues into the conversation. The book includes voices and experiences that have *not* been prioritized in formal policies—for example, women in agriculture, Indigenous women, immigrant women, and women in male-dominated professions. Their contributions highlight the importance of diverse perspectives for a fuller understanding of climate change impacts and ways forward. To place the contributions in context, this introductory chapter briefly traces the history of feminist scholarship in climate change, focusing on key themes in rural and resource contexts of the Global North before introducing the authors and their work.

## Feminist Insights on Gender, Intersectionality, and Climate Change

As early as the 1990s, some feminist scholars were highlighting the gendered nature of natural disasters (e.g., Enarson & Morrow, 1998; Fothergill, 1996). This literature departed from previous disaster research by analyzing gender as more than "a dichotomous survey variable"; rather, it considered the social structures and relations that shaped differential experiences of disaster for women compared to men (Fothergill, 1996, p. 33). At this time, vulnerability was already recognized as a product of social conditions, rather than an inherent characteristic of women or other "vulnerable" groups (Anderson, 1994). The feminist disaster literature would provide an important foundation for later research more explicitly framed by climate change.

Despite this foundational work, little attention was paid to the gender/climate change nexus until the early 2000s. A historical Google Scholar search reveals no sources with titles including "climate change" plus either "gender," "woman," or "women" until 2001—even though the term "climate change" had come into regular usage by the 1970s (NASA, 2008). Replacing "climate change" with "global warming" also produces few results until the mid-2000s. However, beginning in the early 2000s, a body of feminist literature began to link the themes of gender and disaster together with climate change. Key themes in the gender/climate literature to date include disproportionate female fatality during major disasters (Neumayer & Plümper, 2007); climate adaptation and gendered work roles (e.g.,

Alston, 2010; Reed et al., 2014); women's underrepresentation in climate policy and decision making (Denton, 2002; Skutsch, 2002); and the gendered nature of climate change discourse and techno-scientific solutions (e.g., Hultman, 2013; MacGregor, 2010)—to name a few. Gendered vulnerability has also often been linked to the feminization of poverty (e.g., Cannon, 2002).

Identifying the gendered causes and effects of climate change has been foundational for future research and activism. At times, however, the discourse has made overly generalized and universalizing claims about women's and men's different experiences and abilities (Arora-Jonsson, 2011). Certain generalities are common in the discourse—for example, "climate change affects women more than men"; "men's lifestyles contribute to climate change more than women's"; and "women are the most vulnerable to climate change." Such facts, based on high-level aggregation of studies and linked to widespread structures of inequality (e.g., feminization of poverty) or privilege, have helped to draw attention to the gendered dimensions of climate change. Yet, like most other soundbites, they also tend to oversimplify the issues.

Culture, race, socio-economic class, sexuality, age, and ability can create dramatic differences between those in the group "women" and between those in the group "men." Accordingly, in the 2010s, attention turned to intersectionality as a useful framework for better understanding people's context-specific experiences of climate change. An intersectional approach supports careful analysis of the complex causes and impacts of climate change. It maintains an explicitly feminist approach while avoiding homogenizing categories or universal claims.

Intersectional analysis can be traced back to the work of early African-American feminists, such as Sojourner Truth in the 1800s and the Combahee River Collective's 1977 Black Feminist Statement. The term "intersectionality" was itself coined by legal scholar Kimberlé Crenshaw (Crenshaw, 1989, 1991), who critiqued anti-discrimination policies for focusing separately on individual axes of identity, such as gender *or* race, but not the unique experiences created by their intersection. An intersectional approach has the potential to reveal both contextual nuances and deeper structural factors that perpetuate inequality and, by extension, create vulnerability to climate change. Intersectionality is multi-level: It can capture both identity-based experiences *and* reveal persistent systemic and structural issues, such as patriarchy and colonialism. Such analyses hold the potential to inform policy at the local level and beyond, potentially leading to transformation of social relations in the face of climatic threats.

The literature on climate change and intersectionality has followed a similar trajectory to that on gender and climate change—that is, it emerged primarily from disaster research. The starkly gendered and racialized impacts of Hurricane Katrina prompted several important applications of intersectionality (e.g., Pyles & Lewis, 2010; Weber & Hilfinger Messias, 2012). For example, Loretta Pyles and Judith Lewis (2010) challenged the discourse of Hurricane Katrina as a "great equalizer" by noting differences in how African-American and White women experienced the storm, and how they used their agency differently to respond. In the same volume, Libertad Chavez Rodriguez (2010, pp. 70–71) noted that, "not every woman has the same degree of vulnerability. Not all women are in a position worse off than men during disasters." Although aggregated gender trends are important, they do not directly translate to a universal experience.

Researchers have increasingly called for more intersectional work on climate change, noting in particular the scarcity of *empirical* intersectional research (Moosa & Tuana, 2014; Thompson-Hall et al., 2016; Walker et al., 2019). Through our own work, we have attempted to develop intersectional frameworks for empirical research on environment and climate, drawing particular attention to individual variables and causal structures, and to provide conceptual maps and sensitizing questions for multi-level analyses of hazard events (Fletcher, 2018; Walker et al., 2019). Alongside considerations of identity, experience, and agency, these frameworks also consider such structural aspects as power, ideology, and (re)production. Each of the chapters in this book contributes to one or more components of an intersectional approach—from work and bodily experience to discursive representation and political power.

## Gender and Climate Change: Rural and Resource Contexts in the Global North

Much of the scholarly and governmental literature on gender and climate change has focused on the Global South. This focus is important and certainly justified, considering how long histories of colonialism and imperialism, ongoing western hegemony, and persistent geopolitical inequality have constructed conditions of vulnerability in these contexts. Higher-income countries also hold significant responsibility for causing climate change. In the Global South, factors like poverty, livelihood precarity, and weak institutional supports mean that the impacts of climate hazards may be strongly and immediately felt (e.g., Bob & Babugura, 2014). For example, in rural areas within poor countries, climate change can have direct and profound impacts on food security (Mbuli et al., 2021).

Relatively little attention has been given to the gender dimensions of climate change in wealthier countries of the Global North—perhaps because their higher average incomes, insurance, and social safety nets can help buffer the most severe impacts of climate hazards. In 2017, one of the first edited collections solely focused on gender and climate change in the Global North appeared: Marjorie Griffin Cohen's *Climate Change and Gender in Rich Countries: Work, Public Policy and Action* brought a gender lens to such issues as construction, transportation, and activism. Even more recently, Gunnhildur Lily Magnusdottir and Annica Kronsell released their 2021 edited collection, *Gender, Intersectionality, and Climate Institutions in Industrialised States*. Both volumes provide important insights from industrialized countries, with particular emphasis on policy and climate action. Both books include chapters on rural and/or resource contexts, although this is not their main focus. But, as Griffin Cohen (2017) notes in her introduction, countries in the Global North may have much to learn from research and climate initiatives in the Global South—particularly when it comes to gender considerations.

Despite the relatively privileged position of wealthy countries on the global stage, power issues *within* countries like Canada, Spain, and Australia have resulted in highly differentiated climate impacts, including differences between rural and urban settings. In these countries, rural and resource-dependent communities face unique socio-economic challenges, including depopulation, the effects of neoliberal policy regimes, and boom-and-bust economic cycles. Rural livelihoods are shaped by specific, historically engrained gender roles, expectations, and ideologies (Buchanan et al., 2016; Fletcher & Knuttila, 2016; Reed et al., 2014; Wilmer & Fernández-Giménez, 2016). Together, these structural challenges and social factors in the Global North influence adaptation strategies, risk assessment, and individual and community resilience.

This is not to suggest that urban people are unaffected by economic and social structures, but these structures manifest quite differently for rural people. For example, access to emergency support is challenging for those living an hour or more away from the nearest firefighting service. In some rural places, depopulation and urban migration have negatively affected access to formal services, like healthcare and education, as well as informal "neighbour to neighbour" supports (Buck-McFadyen et al., 2019; Fletcher et al., 2020).

Indeed, rural communities exist in a complex relationship with climate change. Across the Global North, rural people make important contributions to national economies through food production, trade, and extractive industries like oil, gas, and mining. These sectors are significant contributors to climate change, but they are also dramatically affected by it. Agriculture is particularly vulnerable to the vagaries of climate change, the burdens of which are often felt by individual farmers and ranchers (Fennell et al., 2016). For communities that are heavily dependent on single sectors, an extreme climate event can be devastating for the rural economy as a whole. At the same time, however, rural people hold important local knowledge gleaned from generations of adapting to climate extremes (Warren & Diaz, 2012).

The economy/environment interface is challenging for rural residents, as conflicting material realities of climate change and resource-based livelihoods are reinforced by strongly embedded identities and discourses. For example, oil, gas, mining, and agriculture are male-dominated industries tied to masculine identities (Letourneau & Davidson, this volume). Job losses, therefore, can be both economic and deeply personal (Räthzel & Uzzell, 2011), striking at the heart of one's sense of self (Reed, 2003). An intersectional approach reminds us that the impacts of climate change and the measures we take to avoid them will have significant, and different, ramifications depending on identity. For example, the experience of agricultural drought interacts with farmers' strongly rooted identities, affecting the mental health of rural men as well as their partners, who may also be experiencing heavy workload and/or caregiving responsibilities (Alston et al., this volume; Alston, 2012; Alston & Kent, 2008; Fletcher & Knuttila, 2016).

Intersectional approaches also sensitize us to the operation of power and privilege in rural places. Individual, micro-level, or local experiences and knowledge emerge from widespread and perseverant social structures. For example,

Indigenous, racialized, and Euro-Canadian women in North America experience some similar forms of gender-based inequality—such as lower median incomes compared to men, although both Indigenous and racialized women often fare worse than Euro-Canadian women. Many Euro-Canadian women have also benefited from Canada's colonial history that violently dispossessed Indigenous people of land and resources (Vinyeta et al., 2015). This history has resulted in significant economic, social, and cultural privilege held by White women, in aggregate, compared to Indigenous people in Canada today (Vinyeta et al., 2015). Consequently, we cannot expect the experiences of climate change to be the same for Indigenous and Euro-Canadian women, or for racialized women in general (either "newly immigrated" or longstanding citizens). Some of these differences are illustrated in the chapters that follow—for example, in the contributions from Lakhina and Eriksen, and from Levac and colleagues. The chapters in this volume demonstrate how climate change is felt and known differently depending on our location within social structures of inequality. Our social location, in turn, shapes how we view those structures—the parts we see and the parts we don't see.

## The Collection

This book examines the complex interconnection of gender, intersecting inequalities, and climate change in rural and resource contexts. The chapters focus in detail on three middle-power countries in the Global North[1]—Australia, Canada, and Spain—all of which are simultaneously responsible for, and affected by, the extreme effects of climate change. Together, the authors' contributions reveal the value of feminist, intersectional research to strengthen theoretical understanding of social dimensions of climate change. They provide practical suggestions/recommendations for transformative policy and practice. Reflections on each chapter, provided by authors with lived experience and/or practitioner expertise on the topic, remind us of the deeply personal impacts when hazards hit, and the value of experiential and embodied knowledge to our understanding of climate change and the actions chosen to address it.

In the first chapter, Heidi Walker unpacks the key features of an intersectional approach through her analysis of a major wildfire event in Saskatchewan, Canada. Her contribution is a novel example of applied, empirical intersectional research on climate hazards in the Global North. Walker's work powerfully demonstrates how the intersection of gender and culture shapes not only how people are affected by a wildfire disaster, but also whose knowledge and skills "matter" in formal policy and planning. The chapter addresses themes of representation, agency, and materiality, revealing how climate change affects—and will continue to affect—the places and spaces that mean the most to us.

Nancy Lafleur, an Indigenous author and educator from northern Saskatchewan, experienced first-hand the evacuation procedures described by Walker. In her poignant commentary, Lafleur reveals how strict adherence to policy and procedure, so often designed from a colonial standpoint, may stand in the way of what people truly need to recover. Her narrative teaches us that more effective wildfire evacuation is that which considers cultural and spiritual values, responding to a wide range of needs beyond merely the physical. Lafleur's own actions during the wildfire exemplify the important efforts made by Indigenous people and communities in developing more culturally safe, flexible, compassionate, and responsive evacuation services.

The second chapter takes us to rural Australia, where Shefali Juneja Lakhina and Christine Eriksen explore the situation of climate disaster through the eyes of refugee women who have resettled in hazard-prone regions. The authors describe not only the structural marginalization the women encountered, but also how women deploy their own knowledge to respond and thrive. Their adaptation strategies include building communities of healing and using cultural practices that foster food security.

As a practitioner in the areas of humanitarian work, refugee resettlement, and emergency response, Sherryl Reddy supports and extends Lakhina and Eriksen's analysis. Reddy identifies current shortcomings in institutional responses to settlement and disaster response. Despite organizations' stated commitments to diversity and inclusion, the perspectives of people from refugee backgrounds are still not positioned as fully equal in organizational processes and practices. Even within community engagement efforts, privileged voices may dominate. Despite positive rhetoric, current efforts often fail to fundamentally transform power structures and hierarchies of knowledge. Reddy provides several useful examples of how a truly intersectional and transformative approach can operate and, with the right amount of support, can thrive.

In Chapter 3, Leah Levac, Jane Stinson, and Deborah Stienstra also address the importance of engaging community perspectives for change. They draw on their work with Indigenous and settler women in northern Canada to present guiding principles for effective intersectional impact assessment. The authors connect concepts of inclusion and agency to inform larger lessons about extractivism and climate change, and how critical community-engaged research can highlight key dimensions of environmental problems that might otherwise be overlooked. Their collaborative efforts with a northern community resulted in the creation of a Community Vitality Index, a community-driven data collection tool to help monitor the impacts of extractive industry on the community.

Responding to Levac and colleagues, environmental lawyer Anna Johnston takes a critical perspective to claims about the "new and improved" impact assessment process in Canada. Johnston argues that the process can only be considered an improvement if decision makers are committed to implementation. She points to legal loopholes and points of Ministerial discretion that continue to allow those in power to dodge any serious forms of community engagement. Hence, the opportunity for people in rural communities to use their data, voices, and knowledge to make a real difference in decision-making structures associated with impact assessment remains precarious.

Angeline Letourneau and Debra Davidson, the authors of Chapter 4, discuss the intersection of hegemonic masculinity and environmental discourse in the oil and gas sector of Alberta, Canada. Alberta's oil industry can be viewed as an extreme case study in the intersection of gender, politics, and environmental impacts. They examine how a particular form of masculinity—"petro-masculinity"—manifests in both discourse and practice, reinforcing a highly polarized debate between oil-industry workers and environmental advocates. Applying a multi-level analysis, they demonstrate how the discourse of petro-masculinity extends beyond workers themselves, influencing the (deeply gendered) realm of environmental politics and policy.

Mary Boyden has worked in an extractive industry for most of her life. As one of the first female gold miners in Canada, she has extensive experience in the kind of environments described by Letourneau and Davidson. Through the wisdom shared by her teachers and Elders, Boyden provides an explanation for the kind of hyper-masculine cultures found in extractive industry; she notes the profound imbalance between masculine and feminine found in these work cultures and emphasizes the need for balance. Boyden shares the excellent work she is doing together with Keepers of the Circle—an organization providing support to Indigenous women working in resource industries. Through these efforts and their recent work on impact assessment, women at Keepers of the Circle are creating a more balanced way forward for communities and the environment. Recognizing the power of story and the intrinsic value of knowledge passed orally, Mary's commentary is provided as a conversational interview with one of the book's editors.

Moving on to Chapter 5, Federica Ravera, Elisa Oteros-Rozas, and María Fernández Giménez address the importance of everyday experience in their chapter on women pastoralists experiencing climate change in Spain. Drawing on the concept of embodiment, the authors question the dominant scientific discourse of climate change, with its emphasis on abstracted biophysical and climatological causes and effects. Instead, the authors turn the biophysical emphasis on its head by emphasizing the deeply subjective and embodied experience of climate change on our individual, physical, and gendered bodies. Lived experience affects not only the experience of climate change, but also how people perceive and adapt to it. Highlighting the intersection of gender, age, and geographic location (urban and rural), Ravera and colleagues describe how women from pastoralist backgrounds—particularly young women—are stereotyped and marginalized by urban activists. The authors connect these individual experiences to "structural forms of inequity," which "are reproduced, reinforced, or challenged by socio-cultural norms, ideologies, and dominant discourses." They also illustrate the value of a multi-level analytical lens, explaining how people from different social locations and knowledge traditions have different levels of influence in shaping environmental, and specifically climate, policies, plans, and strategies.

The lived experience of climate change is poignantly articulated in the reflection by one of Ravera and colleagues' own research participants—pastoralist Lucía Cobos, from Spain. In her poetic contribution, Cobos notes the highly interconnected nature of her work and wellbeing with the natural environment. Pastoralists' knowledge is intimately connected to the heat and the rain. For those whose livelihood depends on the land, rain is not just rain, but rather "pennies that fall from the sky." In drought-prone regions facing climate change, Lucia notes, every day we become poorer.

In Chapter 6, Rachel Reimer and Christine Eriksen discuss another group whose livelihood is strongly connected to the land: Those who work in mountain professions like guiding and wildland firefighting. The authors examine the construction of mountain environments as gendered and racialized spaces—as a frontier "wilderness" in which Whiteness and hegemonic forms of masculinity thrive. Their data provide stark examples of the resulting sexism, homophobia, and transphobia in mountain guiding and wildland firefighting. As these occupations become increasingly risky due to climate change, gendered stereotypes encourage risk-taking—sometimes with life-or-death consequences. The chapter notes the ideological linkage of masculinity with competence in mountain and wildland professions and highlights the implications for leadership in the context of high, hazardous places.

In her experiential reflection on Chapter 6, Alison Criscitiello—a mountain guide, former US ranger, glaciologist, and founder of Girls on Ice—further discusses the gendered dimensions of risk-taking in mountain professions. As a young, queer woman on the ice, Criscitiello experienced the masculine construction of mountain professions documented by Reimer and Eriksen. Her lived experience of being underestimated and stereotyped reflects those of many women in male-dominated sectors, where hegemonically masculine characteristics like physical strength and size are normalized and valued. By establishing a mountain leadership program for girls, Criscitiello exemplifies the embodied feminist leadership necessary to create the inclusive mountain cultures described by Reimer and Eriksen.

In the final chapter of this volume, researchers Margaret Alston, Josephine Clarke, and Kerri Whittenbury present important research data on the lives of Australian farm women experiencing drought. In an attempt to deal with this growing climate risk, government authorities implemented water management policies—but with little attention to the needs of the affected communities. Alston and colleagues document the resulting effects: Mental health consequences for farm men and over-extension of farm women's often-invisible physical and emotional labour. Patriarchy, as a social structure, reproduces itself as women's contributions go unrecognized. Their research shows that when policy is constructed without a gender lens, significant social and health problems are exacerbated.

Alana Johnson, an agricultural leader and farmer in Australia, supports their findings. Johnson strongly calls for women's activism and agency to change the status quo of agriculture—particularly in the context of climate stress. She describes, in detail, the enduring structures that subordinate women's work to men's and documents how water scarcity brought by drought exacerbates the financial and physical burdens for women in farming households. She also provides evidence of positive and progressive change by documenting the political initiatives and progress that women have made in the last two decades. The trajectory of these efforts gives evidence for the title of her reflection: "What is Man-Made can be Unmade."

## Valuing Diverse Knowledge Systems in Feminist Climate Change Scholarship

By explicitly bringing together academic, experiential, and practitioner perspectives for each chapter topic, this book draws upon a rich history of feminist epistemology and critiques of science (e.g., Haraway, 1988; Harding, 1993) that have questioned the "image of science as objective and as both interest- and power-free" (Tuana, 2013, p. 24). We demonstrate the value and importance of multiple ways of knowing and being, which shape the experience of both climate change and climate action.

Recognizing the value of multiple perspectives does not necessitate a retreat into ontological relativism, however—and such a retreat would be particularly problematic in the context of climate change. We illustrate an approach that values, respects, and engages with multiple ways of knowing to describe a reality that *exists*—a reality in which the climate is changing, humans (some of us more than others) are causing climate change, and that change does not affect us equally. In a time of global crisis and ongoing structural inequity, it becomes necessary to assert that certain ways of being and acting in the world, such as striving for equity, social and climate justice, and ecological balance, *are* preferable to others, but that only through epistemological diversity can we recognize and hope to achieve those goals.

This collection emphasizes the importance of weaving academic analysis together with important perspectives from different vantage points to produce a novel understanding and point to a broad array of possible actions. Our contributors often agree with each other, but not always. Nonetheless, taken together, their contributions provide us with the privilege of many partial perspectives (Haraway, 1988). Such perspectives reveal the workings of structures, whether ideological, political, or economic, in need of change; knowledge systems, individuals, and social groups that need to be included; and strategies for action that can lead to more equitable climate transformation.

## Note

1 Although the term "Global North" may sound inaccurate when discussing Australia due to its southern location in the world, the term "Global North" refers not "to a geographic region in any traditional sense but rather to the relative power and wealth of countries in distinct parts of the world" (Braff & Nelson, 2022, n.p.). It is acknowledged that socioeconomic and power disparities also exist *within* countries of the global North and, indeed, this understanding is a premise of this book.

## References

Alston, M. (2010). Gender and climate change in Australia. *Journal of Sociology, 47*(1), 53–70. https://doi.org/10.1177/1440783310376848

Alston, M. (2012). Rural male suicide in Australia. *Social Science & Medicine, 74*(4), 515–522. https://doi.org/10.1016/j.socscimed.2010.04.036

Alston, M., & Kent, J. (2008). The Big Dry: The link between rural masculinities and poor health outcomes for farming men. *Journal of Sociology, 44*(2), 133–147. https://doi.org/10.1177/1440783308089166

Anderson, M. (1994). Understanding the disaster-development continuum: Gender analysis is the essential tool. *Focus on Gender, 2*(1), 7–10.

Arora-Jonsson, S. (2011). Virtue and vulnerability: Discourses on women, gender and climate change. *Global Environmental Change, 21*(2), 744–751. https://doi.org/10.1016/j.gloenvcha.2011.01.005

Australian Disaster Resilience Knowledge Hub. (n.d.). *Black Summer bushfires, NSW, 2019–20*. Retrieved November 18, 2021, from https://knowledge.aidr.org.au/resources/black-summer-bushfires-nsw-2019-20/

Bob, U., & Babugura, A. (2014). Contextualising and conceptualising gender and climate change in Africa. *Agenda, 28*(3), 3–15. https://doi.org/10.1080/10130950.2014.958907

Bonewit, A., & Shreeves, R. (2015). *The Gender Dimension of Climate Justice*. European Parliament, Directorate-General for Internal Policies.

Braff, L., & Nelson, K. (2022). The Global North: Introducing the region. In N. T. Ferndandez & K. Nelson (Eds.), *Gendered lives: Global issues*. State University of New York. https://genderedlives.americananthro.org

Buchanan, A., Reed, M. G., & Lidestav, G. (2016). What's counted as a reindeer herder? Gender and the adaptive capacity of Sami reindeer herding communities in Sweden. *Ambio, 45*(S3), 352–362. https://doi.org/10.1007/s13280-016-0834-1

Buck-McFadyen, E., Isaacs, S., Strachan, P., Akhtar-Danesh, N., & Valaitis, R. (2019). How the rural context influences social capital: Experiences in two Ontario communities. *Journal of Rural and Community Development, 14*(1), 1–18.

Cannon, T. (2002). Gender and climate hazards in Bangladesh. *Gender & Development, 10*(2), 45–50. https://doi.org/10.1080/13552070215906

Cohen, M. G. (2017). Why gender matters when dealing with climate change. In M. G. Cohen (Ed.), *Climate change and gender in rich countries: Work, public policy and action* (pp. 3–18). Routledge.

Crenshaw, K. (1989). Demarginalizing the intersection of race and sex: A Black feminist critique of antidiscrimination doctrine, feminist theory, and antiracist politices. *University of Chicago Legal Forum, 1989*(1), 139–167.

Crenshaw, K. (1991). Mapping the margins: Intersectionality, identity politics, and violence against women of color. *Stanford Law Review, 43*(6), 1241–1299. https://doi.org/10.2307/1229039

Deb, P., Moradkhani, H., Abbaszadeh, P., Kiem, A. S., Engström, J., Keellings, D., & Sharma, A. (2020). Causes of the widespread 2019–2020 Australian Bushfire Season. *Earth's Future, 8*(11), 1-17. https://doi.org/10.1029/2020EF001671

Denton, F. (2002). Climate change vulnerability, impacts, and adaptation: Why does gender matter? *Gender & Development, 10*(2), 10–20. https://doi.org/10.1080/13552070215903

Di Liberto, T. (2021, October 1). *Astounding heat obliterates all-time records across the Pacific Northwest and Western Canada in June 2021*. National Oceanic and Atmospheric Administration. https://www.climate.gov/news-features/event-tracker/astounding-heat-obliterates-all-time-records-across-pacific-northwest

Douty, A. (2021, July 1). *Temperature of 121 F sets new national record high in Canada*. AccuWeather. https://www.accuweather.com/en/weather-news/temperature-of-121-f-sets-new-national-record-high-in-canada/971838

Enarson, E., & Morrow, B. (1998). *The Gendered Terrain of Disaster: Through Women's Eyes*. Praeger.

Fennell, K. M., Jarrett, C. E., Kettler, L. J., Dollman, J., & Turnbull, D. A. (2016). "Watching the bank balance build up then blow away and the rain clouds do the same":

A thematic analysis of South Australian farmers' sources of stress during drought. *Journal of Rural Studies*, *46*, 102–110. https://doi.org/10.1016/j.jrurstud.2016.05.005

Fletcher, A. J. (2018). More than women and men: A framework for gender and intersectionality research on environmental crisis and conflict. In C. Fröhlich, G. Gioli, R. Cremades, & H. Myrttinen (Eds.), *Water Security Across the Gender Divide* (pp. 35–58). Springer International Publishing. https://doi.org/10.1007/978-3-319 -64046-4

Fletcher, A. J., Akwen, N. S., Hurlbert, M., & Diaz, H. P. (2020). "You relied on God and your neighbour to get through it": Social capital and climate change adaptation in the rural Canadian Prairies. *Regional Environmental Change*, *20*(2), 61. https://doi.org/10 .1007/s10113-020-01645-2

Fletcher, A. J., & Knuttila, E. (2016). Gendering change: Canadian farm women respond to drought. In H. Diaz, M. Hurlbert, & J. Warren (Eds.), *Vulnerability and adaptation to drought: The Canadian prairies and South America* (pp. 159–177). University of Calgary Press.

Fothergill, A. (1996). Gender, risk, and disaster. *International Journal of Mass Emergencies and Disasters*, *14*(1), 33–56.

Griffin Cohen, M. (Ed.). (2017). *Climate change and gender in rich countries: Work, public policy and action*. Routledge.

Haraway, D. (1988). Situated knowledges: The science question in feminism and the privilege of partial perspective. *Feminist Studies*, *14*(3), 575–599.

Harding, S. (1993). Rethinking standpoint methodology: What is "strong objectivity"? In L. Alcoff & E. Potter (Eds.), *Feminist epistemologies* (pp. 49–82). Routledge.

Hultman, M. (2013). The making of an environmental hero: A history of ecomodern masculinity, fuel cells and Arnold Schwarzenegger. *Environmental Humanities*, *2*, 83–103.

MacGregor, S. (2010). A stranger silence still: The need for feminist social research on climate change. *The Sociological Review*, *57*, 124–140. https://doi.org/10.1111/j.1467 -954X.2010.01889.x

Magnusdottir, G. L., & Kronsell, A. (Eds.) (2021). *Gender, intersectionality and climate institutions in industrialised states*. Routledge. https://doi.org/10.4324/9781003052821

Mbuli, C. S., Fonjong, L. N., & Fletcher, A. J. (2021). Climate change and small farmers' vulnerability to food insecurity in Cameroon. *Sustainability*, *13*(3), 1523. https://doi .org/10.3390/su13031523

Met Office. (2021, November 3). *Climate change drives Europe's record 2021 summer*. Met Office. https://www.metoffice.gov.uk/about-us/press-office/news/weather-and-climate /2021/2021-european-summer-temperature-impossible-without-climate-change

Moosa, C. S., & Tuana, N. (2014). Mapping a research agenda concerning gender and climate change: A review of the literature. *Hypatia*, *29*(3), 677–694. https://doi.org/10 .1111/hypa.12085

NASA. (2008). *Whats in a Name? Global Warming vs. Climate Change*. https://www.nasa .gov/topics/earth/features/climate_by_any_other_name.html

Neumayer, E., & Plümper, T. (2007). The gendered nature of natural disasters: The impact of catastrophic events on the gender gap in life expectancy, 1981–2002. *Annals of the Association of American Geographers*, *97*(3), 551–566. https://doi.org/10.1111/j.1467 -8306.2007.00563.x

Pyles, L., & Lewis, J. (2010). Women, intersectionality, and resistance: In the context of Hurricane Katrina. In S. Dasgupta, İ. Şiriner, & P. Sarathi De (Eds.), *Women's encounter with disaster* (pp. 77–86). Frontpage Publications Ltd.

Räthzel, N., & Uzzell, D. (2011). Trade unions and climate change: The jobs versus environment dilemma. *Global Environmental Change, 21*(4), 1215–1223. https://doi .org/10.1016/j.gloenvcha.2011.07.010

Reed, M. G. (2003). *Taking stands: Gender and the sustainability of rural communities.* UBC Press. https://www.ubcpress.ca/taking-stands

Reed, M. G., Scott, A., Natcher, D., & Johnston, M. (2014). Linking gender, climate change, adaptive capacity and forest-based communities in Canada. *Canadian Journal of Forest Research, 44*, 995–1004. https://doi.org/10.1139/cjfr-2014-0174

Rodriguez, L. C. (2010). Gender-biased social vulnerability on disasters and intersectionality. In S. Dasgupta, İ. Şiriner, & P. Sarathi De (Eds.), *Women's encounter with disaster* (pp. 62–76). Frontpage Publications Ltd.

Seager, J. (2014). Disasters are gendered: What's new? In A. Singh & Z. Zommers (Eds.), *Reducing disaster: Early warning systems For climate change* (pp. 265–281). Springer.

Skutsch, M. M. (2002). Protocols, treaties, and action: The "climate change process" viewed through gender spectacles. *Gender & Development, 10*(2), 30–39. https://doi .org/10.1080/13552070215908

Thompson-Hall, M., Carr, E. R., & Pascual, U. (2016). Enhancing and expanding intersectional research for climate change adaptation in agrarian settings. *Ambio, 45*(S3), 373–382. https://doi.org/10.1007/s13280-016-0827-0

Tuana, N. (2013). Gendering climate knowledge for justice: Catalyzing a new research agenda. In M. Alston & K. Whittenbury (Eds.), *Research, action and policy: Addressing the gendered impacts of climate change* (pp. 17–31). Springer. http://www.springer .com/environment/global+change+-+climate+change/book/978-94-007-5517-8

Vinyeta, K., Powys Whyte, K., & Lynn, K. (2015). *Climate change through an intersectional lens: Gendered vulnerability and resilience in Indigenous communities in the United States* (PNW-GTR-923; p. PNW-GTR-923). U.S. Department of Agriculture, Forest Service, Pacific Northwest Research Station. https://doi.org/10.2737/PNW-GTR-923

Walker, H. M., Culham, A., Fletcher, A. J., & Reed, M. G. (2019). Social dimensions of climate hazards in rural communities of the global North: An intersectionality framework. *Journal of Rural Studies, 72*, 1–10. https://doi.org/10.1016/j.jrurstud.2019 .09.012

Warren, J. W., & Diaz, H. P. (2012). *Defying Palliser: Stories of resilience from the driest region of the Canadian prairies.* Canadian Plains Research Center Press.

Weber, L., & Hilfinger Messias, D. K. (2012). Mississippi front-line recovery work after Hurricane Katrina: An analysis of the intersections of gender, race, and class in advocacy, power relations, and health. *Social Science & Medicine, 74*(11), 1833–1841. https://doi.org/10.1016/j.socscimed.2011.08.034

Wilmer, H., & Fernández-Giménez, M. E. (2016). Some years you live like a coyote: Gendered practices of cultural resilience in working rangeland landscapes. *Ambio, 45*(S3), 363–372. https://doi.org/10.1007/s13280-016-0835-0

# 1 Wildfire in Northern Saskatchewan

## Reflections for Intersectional Climate Hazards Research and Adaptation Practice

*Heidi Walker*

***Heidi Walker*** *recently completed her PhD in Environment and Sustainability from the University of Saskatchewan, where she used an intersectional lens to explore experiences of, and responses to, wildfire. She currently works as a Research Associate with the Natural Resources Institute at the University of Manitoba. In that role, she supports a project aiming to enhance the integration of qualitative research methods in impact assessment processes. Her main research interests involve the development of equitable and locally responsive approaches to climate change adaptation and impact assessment.*

## Introduction

Wildfire has and will continue to impact communities in the boreal forest regions of Canada, especially as climate change contributes to increases in fire occurrence and intensity over the coming decades (Wotton et al., 2017). In the province of Saskatchewan, climate change is expected to contribute up to a two-fold increase in forest area burned by mid-century and a three- to five-fold increase by 2100 (Wittrock et al., 2018). While fire is a vital ecological process, more frequent and intense fires, combined with expanding development in wildland–urban interfaces, increase risk for local and Indigenous communities in forested regions (Christianson, 2015; Eriksen, 2014).

In 2015, northern Saskatchewan experienced a particularly extreme wildfire season, which resulted in approximately 1.7 million hectares of forest burned and the largest evacuation in the province's history. The La Ronge tri-community— a socially and jurisdictionally complex region of northern Saskatchewan—was among the most impacted areas of the province. While no fatalities or major infrastructural damage occurred within the community itself, residents experienced the event and were impacted in diverse ways.

DOI: 10.4324/9781003089209-2

Increasingly, research shows that climate change interacts in complex ways with existing social dynamics and inequalities, such as gendered norms and expectations, economic disparities, racism, and colonialism, that shape context-specific forms of privilege and marginalization (Iniesta-Arandia et al., 2016; Tschakert et al., 2013). When climate-related hazards occur, they become layered onto these existing dynamics, resulting in differential experiences of such events—and the responses to them—across lines of gender, "race", ethnicity, age, socio-economic status, and other identity attributes. Examples of these complex outcomes have been demonstrated in the context of drought (Alston, 2010; Fletcher & Knuttila, 2016), severe storms (Seager, 2006; Weber & Hilfinger Messias, 2012), flooding (Culham, 2020), and wildfire (Alston, 2017; Eriksen, 2014; Scharbach & Waldram, 2016; Tyler & Fairbrother, 2013). Such examples suggest that understanding how climate hazards interact with existing social landscapes can help identify the root causes of vulnerability and build more equitable responses to these hazards.

Intersectionality has been promoted as an analytical framework for studying these social dynamics in the climate change context, though there remain few empirical examples of its application in climate hazards research in the Global North (Walker et al., 2019). Rooted in Black feminist thought, intersectionality refers to the "interaction between gender, race and other categories of difference in individual lives, social practices, institutional arrangements, and cultural ideologies and the outcomes of these interactions in terms of power" (Davis, 2008, p. 68). This definition describes some of the key elements of intersectionality. First, it recognizes that identity categories do not operate separately, but intersect and co-constitute one another to create context-specific social locations and experiences of the world (Crenshaw, 1991; Hankivsky, 2014). Second, these categories operate within and across multiple levels of analysis—from the micro-scale (lived experiences and subjectivity formation) to meso- and macro-social structures and symbolic representation (Winker & Degele, 2011). Third, multiple systems of power (e.g., racism, sexism, colonialism) interlock to create context-specific forms of privilege and oppression (Cho et al., 2013; Collins, 2017). Along with the examination of power and its intersectional effects, intersectionality scholars promote attention to expressions of agency that challenge context-specific oppressions (Djoudi et al., 2016; Fletcher, 2018).

This chapter synthesizes some of the insights and lessons learned through an intersectional analysis of the 2015 wildfires in La Ronge, including the time periods during and after the event. It first provides an overview of the case study and methodology. Then, weaving together empirical findings from a media analysis and semi-structured interviews, it discusses methodological insights and practical lessons for emergency response and adaptation vis-à-vis each core element of an intersectional analysis (i.e., intersecting categories; multilevel analysis; power relations; agency). The elements for intersectional analysis are deeply intertwined and so their division in this chapter is a rather false separation; however, it is one way of grappling with the enormous complexity that an intersectional analysis presents. Moreover, I recognize that there is no single methodology for empirical intersectional research and that methods must be adapted to each specific context (Kaijser & Kronsell, 2014). Methodological insights highlighted in this chapter are simply meant to spark discussion about the various ways that intersectionality might be applied within climate hazards research.

## La Ronge Wildfire Case Study

La Ronge is considered a tri-community that includes the municipality of La Ronge, the northern village of Air Ronge, and the Lac La Ronge Indian Band (LLRIB). Each is a distinct jurisdiction with its own elected leadership, but they share a number of key services, such as waste and water management and the local fire service. The tri-community has a total population of approximately 6000, of which approximately 75 percent identify as Indigenous (Statistics Canada, 2016).

In 2015, the leaders of the three communities jointly issued a mandatory evacuation order as wildfires moved nearer to the community limits. An evacuation order was also issued for a number of nearby unincorporated subdivisions. Some residents stayed within their communities to support the emergency response efforts, but the vast majority evacuated to stay with friends, family, or at Red Cross evacuation centres in other Saskatchewan communities and neighbouring provinces. The mandatory evacuation lasted for approximately two weeks between July 4 and 19, but many residents left earlier due to heavy smoke and were away from their homes for upwards of one month. While no major structural damage affected the tri-community itself, several homes and cabins in the unincorporated subdivisions and surrounding areas were lost in the fires.

The research occurred over 16 months in 2018 and 2019 and included a media review and semi-structured interviews, which are the basis of this chapter. The media review was based on 105 articles from two local and two national news outlets and were analyzed using an intersectional critical frame analysis framework (see Walker et al., 2020). Semi-structured interviews were conducted with 44 participants, including 34 residents (21 women and 13 men; 12 Indigenous residents, 19 non-Indigenous residents, two with blended heritage, and one with unknown ethnic origin) and 10 municipal, First Nation, and provincial government representatives (see Walker et al., 2021a, b). The interview data were transcribed verbatim and both the interview and media data were coded using the qualitative data software package NVivo (QSR, 2018). This chapter synthesizes some of the findings from these studies as they relate to insights for intersectional research and emergency management and adaptation practice. Brief excerpts from the interview and media data are provided to accentuate the examples provided.

Though the original concept for this study originated outside of the community, my approach to this research was informed by principles of ethical feminist community-engaged research that emphasizes reflexivity, respect, and reciprocity (Creese & Frisby, 2011). I acknowledge that my identity as a woman and a White settler-Canadian of European heritage has informed how I approach research, how others perceive me in the research context, and how I am positioned within the macro-scale structures that this intersectional study set out to interrogate

(Hankivsky, 2014). At times while conducting the research—perhaps because I do not fit the dominant imaginary of an academic researcher—I was perceived and explicitly referred to as "the young lady". Simultaneously, as a White settler researcher, I occupied a position of privilege and worked under the inherited baggage of a historical colonial research regime that often took an extractive approach to research with little reciprocal benefit to northern and First Nation communities (Castleden et al., 2008). I attempted to work in opposition to this form of research by prioritizing principles of relationship, respect, and reciprocity. For example, several site visits were made during research development, which informed the research questions and methods chosen. I also developed a collaborative research agreement with the LLRIB, lived in the tri-community for 16 months to develop relationships and enhance my understanding of the local context, hired an undergraduate student research assistant from the LLRIB, volunteered with local organizations, and ensured childcare funds were provided where interested residents would have otherwise been unable to participate in the project. During my time in La Ronge, I attended many local cultural events and am incredibly grateful and humbled by the welcome I received and the patience with which I was taught. In the sections that follow, I present empirical findings from the media and interview data organized thematically by five core elements of intersectional analysis and reflect on their implications for climate hazards research and practice.

## Reflections for Climate Hazards Research and Practice

### *Intersecting Categories of Identity*

Intersectionality recognizes that multiple identity categories are not additive, but co-constitute one another to create context-specific experiences of privilege and marginalization (Davis, 2008; Hankivsky, 2014). Methodologically, the choice of which—and how many—categories of analysis to include remains a significant challenge of an intersectional approach (Iniesta-Arandia et al., 2016; Kaijser & Kronsell, 2014). Some scholars argue that the most relevant identity categories should emerge during fieldwork (Hackfort & Burchardt, 2018; Hankivsky, 2014), while others choose to focus on the experiences of specific social groups (e.g., Bowleg, 2008; Prior & Heinämäki, 2017). A blended approach was taken in this research by applying a sampling strategy that first stratified participants by jurisdictional boundary and then, due to contextual knowledge of historical colonization in the region and gender-differentiated experiences of hazards, ensured the participation of both men and women and Indigenous and non-Indigenous residents within each jurisdictional group. At each interview, I administered a demographic checklist that included other identity categories, such as age, length of residency, and income. This approach—focused on a small set of identity factors based on contextual knowledge of the study location—provided a manageable scope for the intersectional analysis, while also being flexible enough to allow other relevant social attributes such as age and family status to emerge through the fieldwork and data analysis. Due in part to the strategic decisions about participant

sampling, three factors—location, gender, and ethnicity—emerged most promi-
nently as relevant identity attributes in the analysis.

Ultimately, the purpose of intersectional analysis is not to identify the vast
number of categories that contribute to people's experiences of the world, but
those that are most salient in a specific context (Kaijser & Kronsell, 2014). A
limitation of our approach was that by foregrounding certain identity attributes,
the experiences and knowledges of other populations, such as LGBTQ and Two-
Spirit communities, recent immigrants, and people with disabilities, were likely
obscured. Trade-offs may be necessary, but is important to constantly reflect on
whose knowledge and experience may be excluded and the possible implications
it may have for research and practice.

In La Ronge, numerous examples of how identity categories intersected to
create differential experiences of the wildfire event and evacuation were evident.
Residents, for instance, frequently cited impacts to mental and emotional wellbeing, which were experienced in diverse ways across intersections of gender,
race, age, family status, and location (see Walker et al., 2021a). In one example,
an Indigenous woman spoke of the anxiety she experienced while staying in an
urban evacuation host community due to her fears that Social Services would be
called: "…if my kids start misbehaving, are they going to call [the authorities]
on me?" (woman, LLRIB). This experience was, in part, shaped by both gender
roles and expectations where women are more likely than men to take on caregiving roles during emergency events (Whittaker et al., 2016), as well as the lasting legacies of colonization, assimilative policies (e.g., residential schools), and
systemic racism that mean, even today, Indigenous families are more likely than
non-Indigenous families to be investigated by child welfare services (McKenzie
et al., 2016). Gender and race interlocked to influence this particular experience
of evacuation, which is qualitatively different from how non-Indigenous women
with young families experienced the event.

Findings also revealed that even generally privileged segments of society (e.g.,
White, middle-aged, affluent men) were vulnerable to wildfire impacts in particular ways (see also Eriksen & Simon, 2017). For example, a group of residents
who stayed to protect property in a relatively affluent unincorporated subdivision reported feeling confident in their ability to conduct this work because of
their previous firefighting experience and access to resources (pumps, boats)
(see Walker et al., 2021a). They also generally reported feeling better off than
those who evacuated because they were able to make tangible contributions to
firefighting efforts. Despite this, a female participant indicated that a group of
middle-aged men in the community experienced impacts to mental and emotional
well-being, but were unlikely to seek out mental health resources following the
event due, in part, to gendered expectations and norms resulting in stigma around
mental health issues: "It is easier for a woman to say, 'I am having trouble' than
a man, quite frankly—and certainly that age group of men" (woman, unincorporated subdivision).

An intersectional approach provided insight into the diversity of experiences and responses to the wildfire event across intersections of identity. This

nuanced perspective helps to overcome assumptions that entire communities or social groups (e.g., Indigenous communities, women) are uniformly vulnerable or resilient to the effects of climate change (Arora-Jonsson, 2011; Haalboom & Natcher, 2012; Moosa & Tuana, 2014). It also highlights how privilege and vulnerability can occur simultaneously depending on the context (Smooth, 2013). The heterogenous and context-specific experiences of hazards suggest that blanket approaches to emergency management and adaptation planning will not be received by all residents in the same way. An intersectional approach, therefore, points to the need to adjust emergency response and adaptation programmes to the needs of various social groups. For example, it might identify specific groups in need of mental health resources or it might be useful in identifying the root causes of vulnerability produced by histories of colonization.

### *Multilevel Analysis*

One critique of empirical intersectional research is its tendency to emphasize the micro-level of analysis, where the differential experiences of everyday life or a specific phenomenon are assessed across intersections of identity, such as gender, race, and socio-economic status (Chaplin et al., 2019). In line with theoretical intersectionality scholarship, intersectional climate change literature now encourages identifying the interlinkages between the micro-level experiences of climate hazards and the meso- and macro-level social structures, discourses, and axes of inequality that shape those experiences (Hackfort & Burchardt, 2018; Kaijser & Kronsell, 2014). This multilevel analysis clarifies *why* differential experiences of hazards occur, providing greater opportunity to address the root causes of risk and vulnerability (Chaplin et al., 2019).

This research took two approaches to multileveled analysis. First, as promoted by Winker and Degele (2011) and Kaijser and Kronsell (2014), intersectional dynamics within and across three levels were examined: Symbolic representation, micro-level experience, and social structures (e.g., meso- and macro-level institutions). Looking across these levels is important, as phenomena and processes at one level can reinforce or challenge inequalities at others (Fletcher, 2018). In this case study, an intersectional critical frame analysis was applied to examine how media representation of the 2015 wildfire event reflected particular intersecting social identities (symbolic representation). Interviews with community residents revealed the diverse ways in which people experienced the fire event (micro-level), and interviews with both community residents and government agency representatives provided insight into the institutions that have implications for diverse social groups within the tri-community (structural level). The structural level includes the formal and informal "'rules of the game' – the rules, norms, and practices – that structure political, social, and economic life" (Chappell & Waylen, 2013, p. 599). Power relationships are deeply embedded within, and influence, interactions between these levels. Examples of how these dynamics manifested in the La Ronge wildfire case study will be provided in the following section ("power relations").

Second, the multilevel analysis was strengthened by applying Bowleg's (2008) insight that people's experiences can provide a window to the macro-scale structures that shape experiences in specific contexts, while knowledge of the sociohistorical context of the study location can bolster the understanding of these connections between micro and macro levels. A challenge with this approach is that researchers who are outsiders to the community often "don't know what we don't know." Living in the community for a significant amount of time and taking a humble learning stance greatly enhanced the understanding of the sociohistorical context and its relationship to residents' experiences of the wildfire event. During interviews, I kept in mind contextual knowledge that may be relevant to the participants' narratives, such as the histories of colonization in northern Saskatchewan and the role of gender relations in shaping experiences of hazards in the Global North.

In some cases, residents' narratives spurred additional learning to better contextualize the related experiences. For example, a man from the LLRIB drew parallels between the large evacuation centres located in urban areas south of La Ronge and experiences of residential school[1]: "I do support [the Red Cross] but they shouldn't have total control. It is like they call it residential school mentality... It is like staying in jail or being in prison." Another participant, revealing intersections with gender expectations, indicated that "the social aspect is humiliating— having to be moved onto cots at the Red Cross in Prince Albert as able-bodied men" (man, unincorporated subdivision).

Learning about the history of residential schools in the study region after these interviews occurred gave these accounts additional depth. The Lac La Ronge (All Saints) Indian Residential School operated from 1907 until it burned down in 1947 and students were transferred to a residential school in Prince Albert, Saskatchewan (Niessen, 2017). Niessen (2017) provides Prince Albert residential school survivors' accounts of their time at the school, including descriptions of dormitory life, harsh punishment, separation from family, and bans on speaking their own language. The residential school in Prince Albert remained open until 1996, which means that many people in La Ronge attended the school themselves or have parents or grandparents who attended. Thus, it is understandable how dormitory-style accommodation at large emergency evacuation centres—especially in Prince Albert—may be reminiscent of residential school experiences. Through this iterative movement between participant experience and sociohistorical context, it became clearer how the legacies of colonization and assimilative policies shape contemporary experience of climate hazards and evacuation (see also McGee et al., 2019; Poole, 2019).

This multilevel analysis may also make visible emergency management and adaptation policy avenues that better address the root causes of risk and vulnerability. For example, during emergency evacuations, those considered most vulnerable (e.g., pregnant women, Elders) are often provided hotel accommodation, rather than having to stay at large evacuation centres. This approach may help lessen the discomfort of the evacuation experience and—particularly for First Nations people—reduce exposure to accommodation arrangements that may be reminiscent of residential school dormitories. However, understanding evacuation experiences within the broader context of the legacies of colonization and assimilative policies raises possibilities for more transformative approaches that work to dismantle these historical inequalities. This includes stronger partnerships with Indigenous communities for emergency and adaptation planning, greater First Nations autonomy over the provision of culturally appropriate evacuation procedures and accommodation, and perceiving Elders as a source of expertise and resilience during emergency events, rather than as a uniformly vulnerable population (Betancur Vesga, 2019; Lambert & Scott, 2019; Poole, 2019). Such approaches respond to intersectionality's call to identify and "address the system that creates power differentials, rather than the symptoms of it" (Djoudi et al., 2016, p. S2498).

### *Power Relations*

Through an intersectional lens, "what makes people vulnerable to climate change, or, alternatively how they experience adaptation and mitigation strategies is the result of multiple factors and processes that are linked together within systems of power" (Hankivsky, 2014, p. 17). These systems of power operate within and across levels of analysis and are informed by intersecting forms of power based on identity, such as racism, colonialism, and sexism (Collins, 2017). The multi-level analysis—as the examples in the previous sections illustrate—can facilitate the identification of how power operates to influence diverse experiences and vulnerabilities to climate hazards.

Power relations can also be identified in empirical research through framings of a phenomenon or event. Framings can influence what social interventions are considered or overlooked (discursive effects), shape how various groups are characterized and how individuals see themselves in relation to an issue or event (subjectification effects), and produce tangible outcomes in people's lives, such as access to resources and decision making (lived/material effects) (Bacchi, 2009). From this perspective, power is not viewed solely as the direct domination of certain individuals or groups by others, but rather as the operation of privilege and exclusion of certain knowledges, assumptions, and values in hazard response and adaptation decision making (Hankivsky, 2014; Kaijser & Kronsell, 2014). The framings of the wildfire event and their effects were identified through both the media review and semi-structured interviews. The media critical frame analysis framework included sensitizing questions related to problem framings, solution framings, and who had voice in shaping these framings (see Walker et al., 2020). The interview data were also broadly organized around these themes as a means of triangulation across methods and data sets.

In terms of *discursive effects*, the media critical frame analysis revealed that wildfire was predominately framed as a threat to immediate physical safety, physical infrastructure, and economic values. The media reports also centred the voices of the actors that responded to these problems (e.g., wildfire and emergency management agencies). Such framings are discursive because they limit how we think about wildfire response and planning, as well as the roles of various actors involved.

The framing of the wildfire event also had *subjectification effects.* The media analysis revealed intersecting dualisms in which women and Indigenous communities were largely represented as in need of care and protection or as passive recipients of external aid. In contrast, male-dominated sectors with key roles in frontline firefighting efforts were often represented as heroes, liberators, and even care providers. For example, one article stated: "The military has been called in to help care for 7,900 people from a northern Saskatchewan community that includes the province's largest First Nation" (CBC News, 2015). Another, which implicitly represented women as in need of defence, read: "Evacuees keeping tabs on the situation from afar cheered the 'eye candy' of young men coming in to defend their home" (Hopper, 2015). These "frontline" actors are, without doubt, important to wildfire response and planning; however, such framings of the event and actors involved also have material implications for diverse social groups.

Interview data revealed *material effects*, where institutional policy and practice reinforced the prioritization of physical impacts and frontline expertise during and after the event. For example, a public engagement session following the fires centred around how the fires had been handled within the community. Those invited to provide background information were from male-dominated sectors, such as wildland and structural firefighting agencies. Women who performed volunteer roles within the community during the fires, such as cooking, delivering meals, cleaning, and caring for pets, felt their contributions were excluded from discussions—a material effect flowing from how the wildfire "problem" was framed. As one woman from La Ronge mentioned, "[…] we just wanted them to be inclusive of everything." Longer-term adaptation strategies have also prioritized the mitigation of physical and economic risk, such as through fuel mitigation treatments. Thus, the emphasis on physical and economic impacts has discursively narrowed the agenda to largely exclude the nonmaterial values and losses experienced by residents and how these losses were experienced in diverse ways across intersections of gender, race, ethnicity, age, location, and socio-economic status.

In some cases, the sidelining of nonmaterial impacts also infused into interpersonal relations during the fires. For example, a woman related the deep sense of loss she experienced after her and her partner's remote cabin burned in the wildfire. It was not just the loss of the physical structure she mourned, but also the memories and relationships that were entangled with the place. She spoke of a fellow community member who "knew we had lost our cabin and he [said] 'I don't get these people that are whining about their losing their cabins when this whole town is being threatened,'" to which she replied, "'whoa okay, but there aren't any houses lost here … and that place has always meant more to us than this place here [primary residence near town]'" (woman, unincorporated subdivision). While many interview participants spoke of the nonmaterial impacts they experienced, this exchange particularly resonated with Cox and colleagues' (2008) study that illustrated, in some cases, the reinforcement of stereotypically masculine modes of coping and the framing of emotional expressions of loss as irrational.

Discourse, institutional policy and practice, subjectivity formation, and the power relations that operate within them both reflect and produce one another (Allan, 2010). In terms of emergency management and adaptation practice, this multilevel analysis of power highlights multiple entry points for positive change. For example, it suggests the need to ensure media representations of wildfire events do not reinforce harmful stereotypes (e.g., of women and Indigenous communities). It also encourages us to examine and challenge assumptions that have resulted in exclusionary institutional arrangements for wildfire response and adaptation, and, as mentioned in the previous section, to consider how to implement more transformative adaptation responses that address historically rooted inequalities and systems of power, such as colonialism. It also reveals the need for participatory engagement processes, where people can give voice to the diverse ways in which they experience, and are impacted by, wildfire (Ajibade & Adams, 2019; Tschakert et al., 2016).

### *Agency, Learning, and Social Change*

While intersectionality promotes the assessment of how power relations contribute to the root causes of risk and vulnerability for diverse social groups, it also encourages attention to how individuals and groups enact their agency to challenge these power relations and build adaptive capacity (Djoudi et al., 2016; Kaijser & Kronsell, 2014). Consideration of agency resists victimization narratives that might otherwise represent entire groups or communities as passively vulnerable to the effect of climate change (Fletcher, 2018).

In the La Ronge wildfire media analysis, a sensitizing question on agency was added to the intersectional critical frame analysis framework—*How are dominant narratives challenged?* (see Walker et al., 2020). Attention to this dimension allowed identification of instances where media reports and residents' perspectives countered the dominant framing of wildfire as primarily a physical and economic problem and of northern and Indigenous residents as passively vulnerable to wildfire. For example, another Indigenous community served as an unofficial host community for LLRIB evacuees, providing cultural, mental, and spiritual resources for its guests (e.g., access to traditional foods, Elders' teachings, cultural ceremonies, and translation services) (Betancur Vesga, 2019). Such alternatives to the "residential school mentality" of mainstream evacuation can help acknowledge needs beyond the physical and financial and, in this case, positioned the First Nation as an active player in the wildfire response. Two women's opinion pieces also provided a much more nuanced perspective of the evacuation experience than that often represented in dominant narratives. One, for example, described ways that individuals enacted their own agency to cope with evacuation and suggested ways that host community residents could meaningfully engage with them (CBC News & Riese, 2016). The other spoke of how she coped with the "evacuation rollercoaster," particularly the mental and emotional tolls of the experience and its residual effects even after residents returned to their communities (Barnes-Connell, 2015).

In the interview data, examples where residents, groups, and communities actively responded to impacts and losses to locally significant values were identified. In some cases, these were responses to physical impacts, such as groups of residents who aided in firefighting efforts and a group of women working at the LLRIB evacuation hub who prepared meals for emergency workers. Revealing an example of how gender and culture intersected to inform agency and to challenge framings of Indigenous women as vulnerable, one LLRIB member said of her own contributions: "that is something I feel is a cultural thing, that would be something my grandmother would do [...] [when you] see somebody working and they are helping you, you do what you can to give back to them."

In many cases, residents' actions were taken in response to losses and impacts to nonmaterial values, such as mental and emotional well-being, sense of place, and cultural identity. For example, the LLRIB Chief and a group of local residents stood at the roadside with a "welcome home" sign as residents re-entered the community after the evacuation, which helped a number of residents regain their sense of place and home: "I thought that was one of the most beautiful things, and I was thinking I really want to return" (man, Air Ronge). A group of women from the local arts council also organized an art exhibition shortly after residents returned, which responded to impacts on mental and emotional well-being and, according to some, was a "very, very valuable healing tool" (woman, unincorporated subdivision).

Despite the myriad of individual and collective actions that contributed to the immediate response and recovery efforts, there was little evidence of action that produced transformative shifts in power relations—though this does not necessarily mean that transformations are not occurring over longer time frames or other societal scales (Heikkinen et al., 2018). Both individual and social learning processes are important to facilitating transformative change that addresses the root causes of risk and vulnerability within adaptation processes (Armitage et al., 2011; Tschakert et al., 2013). In La Ronge, interview participants identified a multitude of individual learning outcomes related to their experiences of the wildfire event, which resulted in changes in perceptions and/or actions for wildfire preparedness. These included instrumental (e.g., technical knowledge about pump use and environmental awareness) and communicative learning outcomes (e.g., ability to work together to achieve common goals, importance of fostering cross-cultural relationships, new perceptions of relationship with the land). The lack of transformative learning and learning outcomes beyond the individual level, which might lead to social change, may be partially due to debriefing and adaptation processes that have emphasized professional expertise over participatory engagement with community members (Walker et al., 2021b). Public engagement sessions during the recovery phase, for instance, focused on the frontline "handling" of the fires and the expertise of male-dominated sectors that responded (e.g., wildland and municipal firefighting, conservation officers). Longer-term adaptation planning focused on fuel mitigation treatments spearheaded by the provincial wildfire management agency and community leaders, with little opportunity for participatory emergency preparedness planning and wildfire risk reduction. As one woman indicated, "I don't feel we are going to be any better prepared for the next time because they are not asking the right questions to the right people" (woman, La Ronge).

The above examples of enacted agency reveal the importance of locally significant values that residents wish to maintain as communities continue to live with fire on the local landscape. Planning efforts should recognize and support these actions, though care should be taken to avoid essentializing certain issues and the groups who respond to them. For example, it is frequently women who respond to nonmaterial impacts, due to gendered ideologies and stereotypes that position them as nurturers and caregivers (Fletcher & Knuttila, 2016). Such issues should not simply be added to adaptation agendas as "women's issues." Rather, the gendered dynamics and power relations that have positioned women as key (and often taken-for-granted) responders and that have resulted in the exclusion of such issues from formal response, recovery, and adaptation processes, should be addressed (Resurrección, 2013).

The lack of transformative outcomes after the 2015 wildfire indicates a need for learning-centred activities that engage the diverse knowledges and experiences of local communities along with external knowledge (Tschakert et al., 2013). In reference to disaster planning, Jacobs argues that

> the knowledge that communities have about their experiences, particularly their experiences of oppression, should be centered in planning processes and outcomes […] it is only through the inclusion of this fundamental knowledge of oppression that the structures of oppression can be addressed, modified and changed.
>
> (Jacobs, 2019, p. 29)

While specific participatory planning techniques are beyond the scope of this chapter, public engagement for both community debriefing following a hazard event and for adaptation planning would benefit from attention to the ideal conditions for individual and social learning, such as critical reflection, collective dialogue, multi-directional knowledge transfer, and trust and relationship building (Armitage et al., 2011; Mezirow, 2012; Tschakert et al., 2013).

## Conclusion

To date, there are few empirical examples of applied intersectional analysis in the study of climate hazards in the Global North. One reason may be a perception that "the concept [of intersectionality] is too abstract for the practical analysis of societal interrelations" (Chaplin et al., 2019, p. 17). This chapter, drawing on the example of a major wildfire event in northern Saskatchewan, has raised one possible approach to make the study of these complex relationships accessible. It fosters the identification of differential experiences of wildfire across intersections of identity attributes, such as gender, race, socio-economic class, and family status. Through multiple methods and multilevel analysis, it also situates these experiences within context-specific structures, discourses, and power relations in recognition that "resilience and vulnerability are not characteristics of social groups but a product of existing societal marginality" (Iniesta-Arandia et al., 2016, p. S391).

This multilevel intersectional analysis provides several insights for building inclusive and transformative emergency and adaptation planning. It suggests that planning efforts should not only respond to the needs of diverse social groups, but also focus on transformative approaches that aim to address the historical and contemporary power relations that constitute the root causes of differential risk and vulnerability. Situating differential experiences of climate hazards within broader social structures and power relationships can be an important first step in identifying transformative adaptation pathways. Transformative approaches also require attention to how local people are actively responding to climate hazards that affect them. Formal planning actions should support and expand these responses, recognizing the locally significant values that residents wish to maintain. Future adaptation planning would also benefit from a participatory, learning-centred approach that underscores diverse locally situated knowledges and experiences and prioritizes the co-production of knowledge between residents and external agencies. By integrating these intersectional insights, emergency and adaptation planning are far more likely to be inclusive, equitable, and effective as communities continue to live with fire in the future.

## Note

1 Residential schools were a key element of what the *Final Report of the Truth and Reconciliation Commission of Canada* refers to as the "cultural genocide" of Indigenous communities in Canada (TRCC, 2015). The schools were for the primary purpose of breaking the link between Indigenous children and their cultural identities. At least 139 residential schools operated across Canada in the 19th and 20th centuries.

## Acknowledgements

This research was funded by the Social Sciences Research Council of Canada (SSHRC). The study was possible because of all those who so generously shared their stories and experiences of the 2015 wildfires.

## References

Ajibade, I., & Adams, E.A. (2019). Planning principles and assessment of transformational adaptation: Towards a refined ethical approach. *Climate and Development*, *11*(10), 850–862.

Allan, E.J. (2010). Feminist poststructuralism meets policy analysis: An overview. In E.J. Allan, S. van Deventer Iverson, & R. Ropers-Huilman (Eds.), *Reconstructing policies in higher education: Feminist poststructural perspectives* (pp. 11–36). Routledge.

Alston, M. (2010). Gender and climate change in Australia. *The Australian Sociological Association, 47*(1), 53–70.

Alston, M. (2017). Gendered outcomes in post-disaster sites. In M.G. Cohen (Ed.), *Climate change and gender in rich countries: Work, public policy and action* (pp. 133–149). Routledge.

Armitage, D., Berkes, F., Dale, A., Kocho-Schellenberg, E., & Patton, E. (2011). Co-management and the co-production of knowledge: Learning to adapt in Canada's Arctic. *Global Environmental Change, 21*(3), 995–1004.

Arora-Jonsson, S. (2011). Virtue and vulnerability: Discourses on women, gender and climate change. *Global Environmental Change, 21*(2), 744–751.

Bacchi, C. (2009). *Analysing policy: What's the problem represented to be?* Pearson Australia.

Barnes-Connell, V.G. (2015, August 6). Evacuation rollercoaster. *The La Ronge Northerner, 61*(28).

Betancur Vesga, S. (2019). *Inside the Rez Cross: An assessment of hosting evacuees during a wildfire disaster in Beardy's & Okemasis First Nation* [Unpublished Master's thesis, University of Saskatchewan]. http://hdl.handle.net/10388/11913

Bowleg, L. (2008). When Black + lesbian + woman ≠ Black lesbian woman: The methodological challenges of qualitative and quantitative intersectionality research. *Sex Roles, 59*(5–6), 312–325.

Castleden, H., Garvin, T., & Huu-ay-aht First Nation. (2008). Modifying Photovoice for community-based participatory Indigenous research. *Social Science & Medicine, 66*(6), 1393–1405. https://doi.org/10.1016/j.socscimed.2007.11.030

CBC News. (2015). *Saskatchewan wildfires force nearly 8,000 people out of homes.* http://cbc.ca/news, accessed 23 July 2020.

CBC News, & Riese, K. (2016, May 5). *Sask. fire evacuee shares experience, advice for Fort McMurray residents.* www.cbc.ca/news/canada/saskatoon/sask-evacuee-fort-mcmurray-1.3569008, accessed 26 March 2021.

Chaplin, D., Twigg, J., & Lovell, E. (2019). *Intersectional approaches to vulnerability reduction and resilience-building.* www.braced.org/resources/i/intersectional-approaches-vulnerability-reduction, accessed 26 July 2020.

Chappell, L., & Waylen, G. (2013). Gender and the hidden life of institutions. *Public Administration, 91*(3), 599–615.

Cho, S., Crenshaw, K., & McCall, L. (2013). Toward a field of intersectionality studies: Theory, applications, and praxis. *Signs, 38*(4), 785–810.

Christianson, A. (2015). Social science research on Indigenous wildfire management in the 21st century and future research needs. *International Journal of Wildland Fire, 24*(2), 190–200.

Collins, P.H. (2017). The difference that power makes: Intersectionality and participatory democracy. *Revista de Investigaciones Feministas, 8*(1), 19–39.

Cox, R.S., Long, B.C., Jones, M.I., & Handler, R.J. (2008). Sequestering of suffering: Critical discourse analysis of natural disaster media coverage. *Journal of Health Psychology, 13*(4), 469–480.

Creese, G., & Frisby, W. (2011). *Feminist community research: Case studies and methodologies.* UBC Press.

Crenshaw, K. (1991). Mapping the margins: Intersectionality, identity politics, and violence against women of color. *Stanford Law Review, 43*(6), 1241–1299.

Culham, A.K. (2020). *The social impacts of flood on the Canadian Prairies* [Unpublished Master's thesis, University of Regina]. http://hdl.handle.net/10294/9322

Davis, K. (2008). Intersectionality as buzzword: A sociology of science perspective on what makes a feminist theory successful. *Feminist Theory, 9*(1), 67–85.

Djoudi, H., Locatelli, B., Vaast, C., Asher, K., Brockhaus, M., & Basnett, B.S. (2016). Beyond dichotomies: Gender and intersecting inequalities in climate change studies. *Ambio, 45*(S3), 248–262.

Eriksen, C. (2014). *Gender and wildfire: Landscapes of uncertainty*. Routledge.

Eriksen, C., & Simon, G. (2017). The affluence–vulnerability interface: Intersecting scales of risk, privilege and disaster. *Environment and Planning A: Economy and Space, 49*(2), 293–313.

Fletcher, A.J. (2018). More than women and men: A framework for gender and intersectionality research on environmental crisis and conflict. In C. Fröhlich, G. Gioli, F. Greco, & R. Cremades (Eds.), *Water security across the gender divide*. Springer International Publishing.

Fletcher, A.J., & Knuttila, E. (2016). Gendering change: Canadian farm women respond to drought. In H. Diaz, M. Hurlbert, & J. Warren (Eds.), *Vulnerability and adaptation to drought: The Canadian Prairies and South America*. University of Calgary Press.

Haalboom, B.J., & Natcher, D.C. (2012). The power and peril of "vulnerability": Lending a cautious eye to community labels in climate change research. *Arctic, 65*(3), 319–327.

Hackfort, S., & Burchardt, H.J. (2018). Analyzing socio-ecological transformations–a relational approach to gender and climate adaptation. *Critical Policy Studies, 12*(2), 169–186.

Hankivsky, O. (2014). *Intersectionality 101*. Simon Fraser University.

Heikkinen, M., Ylä-Anttila, T., & Juhola, S. (2018). Incremental, reformistic or transformational: what kind of change do C40 cities advocate to deal with climate change? *Journal of Environmental Policy & Planning, 21*(1), 90–103.

Hopper, T. (2015, July 8). The battle for La Ronge; Soldiers join fight as flames hit airport, threaten homes. *National Post*, p. A1.

Iniesta-Arandia, I., Ravera, F., Buechler, S., Díaz-Reviriego, I., Fernández-Giménez, M.E., Reed, M.G., Thompson-Hall, M., Wilmer, H., Aregu, L., Cohen, P., Djoudi, H., Lawless, S., Martín-López, B., Smucker, B., Villamor, G.B., & Wangui, E.E. (2016). A synthesis of convergent reflections, tensions and silences in linking gender and global environmental change research. *Ambio, 45*(3), 383–393.

Jacobs, F. (2019). Black feminism and radical planning: new directions for disaster planning research. *Planning Theory, 18*(1), 24–39.

Kaijser, A., & Kronsell, A. (2014). Climate change through the lens of intersectionality. *Environmental Politics, 23*(2), 417–433.

Lambert, S.J., & Scott, J.C. (2019). International disaster risk reduction strategies and Indigenous peoples. *The International Indigenous Policy Journal, 10*(2), 1–21.

McGee, T.K., Mishkeegogamang Ojibway Nation, & Christianson, A.C. (2019). Residents' wildfire evacuation actions in Mishkeegogamang Ojibway Nation, Ontario, Canada. *International Journal of Disaster Risk Reduction, 33*, 266–274.

McKenzie, H. A., Varcoe, C., Browne, A.J., & Day, L. (2016). Disrupting the continuities among residential schools, the sixties Scoop, and child welfare: An analysis of colonial and neocolonial discourses. *The International Indigenous Policy Journal, 7*(2), 1–24.

Mezirow, J. (2012). Learning to think like an adult: Core concepts of transformation theory. In E.W. Taylor & P. Cranton (Eds.), *The handbook of transformative learning: Theory, research, and practice*. Jossey-Bass.

Moosa, C.S., & Tuana, N. (2014). Mapping a research agenda concerning gender and climate change: a review of the literature. *Hypatia, 29*(3), 677–694.

Niessen, S. (2017). *Shattering the silence: The hidden history of Indian Residential Schools in Saskatchewan*, Faculty of Education, University of Regina.

Poole, M.N. (2019). *"Like residential schools all over again": Experiences of emergency evacuation from the Assin'skowitiniwak (Rocky Cree) community of Pelican Narrows* [Master's Thesis, University of Saskatchewan, Saskatoon, Canada].

Prior, T.L., & Heinämäki, L. (2017). The rights and role of indigenous women in the climate change regime. *Arctic Reviews*, *8*, 193–221.

QSR International. (2018). *NVivo qualitative data analysis software, Version 12*. QSR International Pty Ltd.

Resurrección, B.P. (2013). Persistent women and environment linkages in climate change and sustainable development agendas. *Women's Studies International Forum*, *40*, 33–43.

Scharbach, J., & Waldram, J.B. (2016). Asking for a disaster: Being "at risk" in the emergency evacuation of a northern Canadian Aboriginal community. *Human Organization*, *75*(1), 59–70.

Seager, J. (2006). Noticing gender (or not) in disasters. *Geoforum*, *37*(2), 2–3.

Smooth, W.G. (2013). Intersectionality from theoretical framework to policy intervention. In A.R. Wilson (Ed.), *Situating intersectionality: Politics, policy and power*. Palgrave MacMillan.

Statistics Canada. (2016). *Census profile, 2016 Census: La Ronge population center.* Retrieved 28 July 2020 from www12.statcan.ca/census-recensement/2016/dp-pd/prof/details/page.cfm?Lang=E&Geo1=POPC&Code1=1188&Geo2=PR&Code2=47&Data=Count&SearchText=La&SearchType=Begins&SearchPR=01&B1=All&TABID=1

Truth and Reconciliation Commission of Canada (TRCC). (2015). *Final report of the Truth and Reconciliation Commission of Canada: Summary: Honouring the truth, reconciling for the future*. Retrieved 30 July 2020 from www.trc.ca/assets/pdf/Honouring_the_Truth_Reconciling_for_the_Future_July_23_2015.pdf

Tschakert, P., van Oort, B., St. Clair, A.L., & LaMadrid, A. (2013). Inequality and transformation analyses: A complementary lens for addressing vulnerability to climate change. *Climate and Development*, *5*(4), 340–350.

Tschakert, P., Das, P.J., Shrestha Pradhan, N., Machado, M., Lamadrid, A., Buragohain, M., & Hazarika, M.A. (2016). Micropolitics in collective learning spaces for adaptive decision making. *Global Environmental Change*, *40*, 182–194.

Tyler, M., & Fairbrother, P. (2013). Bushfires are "men's business": The importance of gender and rural hegemonic masculinity. *Journal of Rural Studies*, *30*, 110–119.

Walker, H.M., Culham, A., Fletcher, A.J., & Reed, M.G. (2019). Social dimensions of climate hazards in rural communities of the global North: An intersectionality framework. *Journal of Rural Studies*, *72*, 1–10.

Walker, H.M., Reed, M.G., & Fletcher, A.J. (2020). Wildfire in the news media: An intersectional critical frame analysis. *Geoforum*, *144*, 128–137.

Walker, H.M., Reed, M.G., & Fletcher, A.J. (2021a). Applying intersectionality to climate hazards. *Climate Policy*, *21*(2), 171–185.

Walker, H.M., Reed, M.G., & Fletcher, A.J. (2021b). Pathways for inclusive wildfire response and adaptation in northern Saskatchewan. In G. Magnusdottir & A. Kronsell (Eds.), *Gender, intersectionality and climate institutions in industrialised states*. (pp. 226–244). Routledge.

Weber, L., & Hilfinger Messias, D.K. (2012). Mississippi front-line recovery work after Hurricane Katrina: An analysis of the intersections of gender, race, and class in advocacy, power relations, and health. *Social Science & Medicine*, *74*(11), 1833–1841.

Whittaker, J., Eriksen, C., & Haynes, K. (2016). Gendered responses to the 2009 Black Saturday bushfires in Victoria, Australia. *Geographical Research, 54*(2), 203–215.

Winker, G., & Degele, N. (2011). Intersectionality as multi-level analysis: Dealing with social inequality. *European Journal of Women's Studies, 18*(1), 51–66.

Wittrock, V., Halliday, R.A., Corkal, D.R., Johnston, M., Wheaton, E., Lettvenuk, L., Stewart, I., Bonsal, B., & Geremia, M. (2018). *Saskatchewan flood and natural hazard risk assessment*. Saskatchewan Research Council.

Wotton, B.M., Flannigan, M.D., & Marshall, G.A. (2017). Potential climate change impacts on fire intensity and key wildfire suppression thresholds in Canada. *Environmental Research Letters, 12*(9), 1–12.

# Reflection on Chapter 1

## From Point A to Point B

*Nancy Lafleur*

**Nancy Lafleur** *is a Woodland Cree born and raised in northern Saskatchewan. Nancy continues to live in the north where she currently works as a faculty member for the Gabriel Dumont Institute with the Northern Saskatchewan Indigenous Teacher Education Program (NSITEP).*

Another work-year was coming to an end and holiday plans well on their way for summer 2015. I did not anticipate how I would spend those first few weeks, or of the life lessons I would be taught through the experiences, stories, and lives of so many.

I was living and working in Prince Albert—a small city in Saskatchewan, Canada—during the teacher-months and would commute the two-and-a-half-hour drive back to La Ronge on weekends. As the year end was winding down, I had planned on moving back home to La Ronge for the summer and had just finished putting my belongings into storage. The only thing left sitting in my tiny basement-suite was the bed and soon my husband would arrive, and we would complete the transition back home. I knew there was fire activity brewing in the north, but I was confident that the fires would be contained, and I could go on with my life without disaster disruptions. I had many wonderful activities planned for the summer and unfortunately had let my guard down to remembering the impacts forest fires really had on those living in forested areas.

The forest fires in the northern region had summer plans of their own—they were on a collision course with humanity. By the time my husband was on the road from La Ronge to Prince Albert, plans for road closures and possible evacuations were looming; in fact, his travel route was lengthened by a detour caused by a fire crossing the road. He arrived safely much later that afternoon and within 24 hours the north was in full evacuation. Our first thoughts were to figure out how we could best help those who were needing help. We drove to the command centre of the Red Cross, and this is where I truly realized the differences between the two worlds I had been living in.

Within a half hour of being at the command centre, a group came in from a northern community. One of them recognized my husband and immediately came

DOI: 10.4324/9781003089209-3

to ask for assistance. He told Glenn that he was with his elderly mother, and they had just driven in from Saskatoon. His father had just passed away hours prior and they were not allowed to return home. He explained to my husband that they had a place to stay but just needed a few blankets. They required nothing else. As my husband was getting up to go retrieve the blankets, he was told by staff at the command centre that the group would first have to register with the Red Cross before receiving any services. My husband explained that the family did not require services as they had family in Prince Albert and all they required were a few extra blankets. He was told that things were not done that way and forms had to be filled out first and the group would have to be processed.

My husband and I both grew up and continue to live and practise in a culture that gives without asking questions. My grandmother would always remind me that if someone else needed something that I had, and they needed it more than I did, I would have to either lend it or give it to them. Both Glenn and I grew up with this practice in our northern communities. We shared when we could and gave if we had to. My husband, understanding these values, struggled to compromise these teachings when asked to process a group that just needed a few blankets and were not requesting anything else. He understood the challenges the family was facing and the courage it took for them to ask for assistance hours after losing a family member, an Elder. He was compassionate to the stress they were feeling of uncertainty of when they would be allowed home to tend to the funeral and burial of their loved one, or if they would even have a home to return to. He appreciated the need for this family to return home so they could properly grieve and mourn with other family and with their community for the loss of an Elder.

My husband tried to plead with the organization; however, his pleas fell on deaf ears because the policy and guidelines took priority over human compassion for this particular case. My husband, deciding to take matters into his own hands, got up, went to the back and snuck out a few blankets for the family who had just lost a loved one. He gave them his personal phone number and told them if they needed any more blankets, he would look after them personally. After smuggling out the blankets, my husband returned to the front desk where I was doing intake forms and told me that our services could be useful elsewhere. We left.

The following day, all education staff with the organization I worked for were called back to help with the evacuation and we were to report to the Prince Albert Grand Council command centre. Our duties included driving evacuees around the city from point A to point B. We were also given the task of setting up some recreational activities for the children who would need something to do while away from their homes. My husband, an evacuee himself, decided that he would join the transportation group as dispatch.

The first days were filled with chaos and disorganization not realizing what numbers would be flooding into the city to find refuge from the fires. The Prince Albert Grand Council responded to the needs of people with much of the same Cree values I had grown up with. They set up places for people to eat, sleep, and to tend to their hygienic needs. Volunteers came from all walks of life, and donations poured in. The Senator Allan Bird gymnasium was not just a command centre, but it became the place where people came to find friendships, laughter, and to share their stories of how the fire was impacting their lives. For 12 hours a day I drove people and listened to their stories.

One day I drove a mother and her three children to the grocery store. She had a Purchase Order that was very specific on what she could or could not purchase. She was sleeping in a gymnasium full of people and she had four children with her, one that she was looking after as the mother was out on the fire lines. Her husband was also out on the fire lines, fighting fires with many of the other men and women that were not just fighting the fires to save our communities but were doing so to earn some money for their families. On the day I drove her, it was her youngest child's 2nd birthday. It broke my heart that this child could not be at home celebrating her birthday with her family. I decided, while the mother was shopping, to go purchase a birthday cake and some gifts for the child. Upon dropping her off, I gave her the cake and gifts and told her to celebrate her daughter's birthday in the best way she could. She thanked me with tears in her eyes.

I met another young lady that I had transported to the methadone clinic. She told me how she was so afraid of being in the city as it was the city life that almost killed her. She had left for the city at a very early age and quickly became addicted to drugs. She said she had to sell herself to feed her addiction and had lived the street life for many years before returning home to try to beat her addiction. She had just returned back to her home community six months prior to the evacuation and was stressed that her life might have some setbacks in her return to the city. She explained that being surrounded with family supports had been her best therapy and now her family had been separated between Prince Albert and Saskatoon. She explained that she did not get her methadone dose for a few days because the south was slow in preparing for addicts who would arrive, but who would not leave their addictions behind. She told me that there were moments of weakness where she contemplated going back to the familiar streets so she could get drugs. When I asked her what stopped her, she told me that she was with her mother and her child; a child her mother had been caring for. She explained that although much of her extended family was separated, the two people she needed the most were with her, and this was what gave her strength.

There were so many stories such as the few I shared, and I would go home to my tiny basement suite, with just a bed, and thank the Creator for my blessings. Many of the stories and the people in them stayed with me in my dreams, and it took me years to process the dynamics of these lived experiences. As a northerner, I had become so absorbed in the response of helping others that I was not evolved in the social dimensions of its impacts. We were all in survival mode, all doing our best to protect and care for those around us.

About midway into the evacuation, my husband and I were asked to make a delivery to one of the impacted communities. Our commissary load included fresh water, groceries, and dog food. The community we were asked to deliver to was still under high alert as a fire was burning through it and it had already taken the homes of some of the community members. The drive there was like transferring into a movie script and the anticipated unknown could emerge at any turn. It was a 45-minute drive from Prince Albert to Montreal Lake. About 40 minutes down the main highway was a turn that led to the First Nation. We were welcomed by blackened trees and smouldering hot spots that were still coughing up what little life they still had in them. The devastation to the environment was unforgettable; all that was left standing was the blackened trees. Burnt. There was no wildlife, no wild plants, and no signs of any life, except the smell of the charcoal remnants of what once was.

Past the barricades, a new type of living was met with gratitude, and stories. The men and women that greeted our truck and its supplies were beyond any words I could use to describe. They, along with the many that were out fighting the fires, were the real heroes, and for that 20-minute drop-off, they had treated us like heroes for simply dropping off supplies. We had quick exchanges of stories of how the evacuees were coping, and they shared stories of how they were coping with staying back in the midst of the fire in hopes they could save as many homes as possible. Unfortunately, not everyone got to go home to a house that year.

The 2015 experience has provided me with the opportunity to reflect on the cultural paradigms that exist when dealing with a disaster. I observed and participated in a cross-secular response to human need versus family and community need. Although compassion exists within both responses, the need to tend to family resonates more on a spiritual level as opposed to responding on the human and physical level of compassion. The Prince Albert Grand Council definitely responded with heart and tended to everyone's needs without question, conformity, and assignment. The Grand Council organized themselves on the premise of Indigenous values and worldview on family and community cooperation. I understand this concept too well as my experiences with forest fires shaped my ideals around what it means to be a helper within a family and a community.

I grew up mainly in Weyakwin, Saskatchewan. Weyakwin is located in Central Saskatchewan. Growing up, I came to learn and live five seasons: Fall, winter, spring, summer, and fire season. If fire season was active, our small community of Weyakwin became just as active with helicopters flying in and out picking up firefighters. I remember the men in our community lining up on the streets with their sleeping bags and work boots hoping to get picked for the next crew. When trucks filled with both firefighters and their gear would drive around town I knew two things: One—that there was fire that needed attending to, and two—soon there would be money in the community for families. I would hang around outside to see if my uncle Johnny would get picked up. His time out on the fire lines meant spending money for me; growing up with limited funds, this was a big deal. Although I understood the dangers of the fires, somehow, I did not worry too much about my uncle's safety. Somehow, I knew that he would always be safe; and he was.

As a young girl growing up around fires and watching the firefighters go out year after year, I do not ever recall any fatalities associated with fighting the fires. As an adult, I learned to collect this knowledge on how my uncle and the other firefighters stayed safe. My uncle John, like all the others, grew up on the land. They learned to fish, hunt, and survive on it. They also learned to read not just the land, but the animals, weather, and fires. As an adult learner, I came to understand that the success of the firefighters came from traditional teachings of first, respecting the power of fire and, second, to understand the characteristics of fire. They learned to gauge the relationships between fire, land, and weather patterns. As a youth, I had the experience of learning somewhat of these characteristics when a fire broke out close to the community.

I do not recall how the 2015 fire started but do recall that it was situated on the treeline close to the community. There were very few men in the community at that time, as most had been out firefighting and all that were left in the community were women, children, and teens. Within minutes of the fire starting, a group ran to the nearest house, and it was here we started an assembly line of water buckets to try to douse the fire. Thinking that we had beaten the fire, the fire had other plans and resurfaced in the middle of the night. It was at this time I learned that fires travel underground.

Today, 16 July 2021, six years after the 2015 fires, I am sitting in my camper overlooking the lake at Wadin Bay, Saskatchewan; a provincial campsite that is two hours and forty-five minutes north of Prince Albert. We are surrounded by forest fires in every direction possible. We are in the middle of the fire season, and it is fierce. I keep my phone nearby monitoring the latest updates, warnings, and potential emergencies that can arise from the characteristics of the fires. We are at the mercy of the weather; the wind, rain, and lightning will determine the lifeline of fire season, and all we can do is prepare.

How does one prepare for the dynamics of forest fires? Although they are a natural process for rejuvenation and regrowth, the impacts to the human story live vicariously in its outcomes. As a young girl, my grandmother would keep a few bags packed just in case we were evacuated with minutes to spare. We had become immune to this ritual when our community was threatened a few times by wildfires. Living in a fire season paradigm became a state of mind and although humanity can never really control the outcomes and may be able to control physical preparedness, the difficulty lies in the social, emotional, and spiritual response. Wondering where the fire will move, how fast it will move, and how far it will reach is draining so much from the human capacity. When a fire is near a community, imagining the losses can be devastating; to lose the only home one might have, pets, and the vegetation that surrounds it is enduring on the mind and spirit. Although not all fires are a danger to communities, they can prove destructive to traditional lands where families for generations have hunted, fished, collected medicines, and harvested berries. Damage to these spaces of historic and cultural significance is equally distressing.

The fires burning today have many people on edge and, because of social media, tensions are intensifying. I have walked to the edge of the lake to look out at the smoke plumes and wondered what the fire has taken in its path. I have been visited by a small bear and my first thoughts are, "did he flee from the fire?" Again, like always, as a northerner, I am at the mercy of these fires and with really no place to flee. Again, like 2015, fires have divided the province into impact zones and receiver zones. A Facebook post that is warning people to stay out of the smoke reads, "can't they just put them out?" The fires are causing smoke issues for the south. This comment is evident of just how much one can either know about forest fires and how much one does not know. The roads are closed for travel on Highway #2 to La Ronge while three fires cross the road. What will this mean? I do not know; all I can do is prepare myself for a possible power outage and possible evacuation. I just pray that the rain comes as predicted tomorrow so that I do not have to move once again from Point A to Point B.

# 2   Seeking Safe Refuge in Regional Australia

## Experiences of Hazards and Practices of Safety among Women from Refugee Backgrounds

*Shefali Juneja Lakhina and Christine Eriksen*

**Shefali Juneja Lakhina** *is co-founder of Wonder Labs, a California-based social enterprise working to catalyze innovations with communities on the frontline of climate impacts. Since 2005, Shefali has contributed to a range of innovations in disaster risk reduction policy, programmes, and research. She has lived and worked in South and South East Asia, the Middle East and North Africa, South East Europe, Australia, and the United States. Shefali recently founded the Reimagining 2025: Living with Fire Design Challenge to centre the voices of students and early career researchers in wildfire risk reduction efforts.*

**Christine Eriksen** *is Senior Researcher in the Center for Security Studies at ETH Zürich, Switzerland. With a focus on social dimensions of disasters, she gained international research recognition by bringing human geography, social justice, and wildfires into dialogue. Christine is the author of two books and over 75 articles and book chapters, which examine social vulnerability and risk adaptation in the context of environmental history, natural hazards, cultural norms, and political agendas.*

## Introduction

In this chapter, we foreground diverse narratives and experiences of climate change. We seek to problematize notions of *who* belongs in the Australian landscape and *whose* narratives, experiences, and practices count, as Australia reckons with the extreme cascading impacts of the 21st century's climate crisis. We ask, who experiences the variegated impacts of climate change across the Australian landscape? How do climate mobilities contribute to diverse environmentalisms in the Australian landscape? Specifically, how do newly arrived and recently settled

DOI: 10.4324/9781003089209-4

families, women-headed households, and emerging communities from refugee and migrant backgrounds adapt to unfamiliar landscapes? In asking these questions, we seek to extend debates around the politics (Giddens, 2009) and sociology of climate change (MacGregor, 2010), to pursue *social geographies of climate resilience*, to understand who is adapting to climate change, where, and how.

Climate and disaster policy have long acknowledged that age, gender, race, and ethnic identities can result in differential vulnerabilities to climate impacts (United Nations [UN], 1994, 2005, 2015). Over the past decades, disaster research has concretely shown why the social and gendered dimensions of hazards are important to consider (Enarson, 1998; Quarantelli, 1992; Wisner, 1998; Wisner et al., 1994). More recent disaster research has emphasized why this kind of intersectional analysis can be especially important in regional and rural contexts where a range of social, economic, and cultural characteristics can intersect in particular ways to shape people's experiences of climate impacts (Alston, 2011; Djoudi et al., 2016; Eriksen, 2014; MacGregor, 2010; Reed et al., 2014; Whittaker et al., 2016). However, intersectional analysis is still approached as a largely conceptual concern in climate change research and policy (Kaijser & Kronsell, 2014), and it remains generally lacking with regard to regional and rural contexts across the Global North (Walker et al., 2019). There remains a general disconnect between the extant theoretical framings of intersectionality and its integration in empirically grounded research, policy, and practice (Walker et al., 2019). In this chapter, we seek to bridge this gap by presenting empirical evidence of how an intersectional lens, with specific reference to gender, culture, and immigration status, can contribute to grounding disaster research, policy, and practice in an understanding of people's everyday practices of safety during the 21st century's unfolding climate crisis.

In acknowledging the diverse origins, forms, and emerging futures of "intersectionality," including radical queer (Gumbs et al., 2016; Moraga & Anzaldua, 1981), subaltern (Das, 1995; Guha, 1982; Mohanty, 1991; Spivak, 1988), ecofeminism (Agarwal, 1992; Shiva, 1989), Black feminism (Collins, 1990, 2015; Crenshaw, 1991; hooks, 1990, 2000), ecomaternalism (MacGregor, 2006; Williams, 2019), Southern (Connell, 2007), and ecowomanism (Harris, 2016), we focus on "what intersectionality *does* rather than what intersectionality is" (Cho et al., 2013, p. 795). Through this framing, intersectionality enables us to concretely engage with the transformative micropolitics of everyday life, not simply provide a description of people's experiences of climate-related hazards (Collins & Bilge, 2020). In undertaking this kind of intersectional analysis, we join with broader calls to decolonise geographical knowledge (Baldwin, 2017; Clement, 2019), and make space for refugee experiences and voices (Gill, 2018), especially within an emerging analysis of environmental justice (Jacobs, 2019; Ryder, 2017) and climate mobilities (Boas et al., 2019; Fiddian-Qasmiyeh, 2020). While climate change has been identified as a main cause of forced displacement and great migration, a more nuanced and intersectional understanding of climate mobilities remains scarce (Boas et al., 2019; Kelman, 2020). We respond to calls for an intersectional research agenda on climate mobilities (Boas et al., 2019) with a view to develop a more complete understanding of how women's experiences of climate-related hazards are shaped by their mobilities across changing landscapes (Kaijser & Kronsell, 2014; Lama et al., 2020).

In the next section we critically examine recent research that has sought to examine the lived experiences of people from diverse backgrounds who have migrated to Australia in recent decades. We question why refugee experiences remain marginal in both research and policy constructions of climate change impacts and adaptation practices. In the third section, we outline our particular method of intersectional analysis, based on the lead author's doctoral research (2016–2019) in the Illawarra region of New South Wales (NSW), Australia. We demonstrate how an intersectional analysis helped reveal the lived experiences and everyday practices of safety among women, particularly mothers, from diverse refugee backgrounds who resettled in the Illawarra between 2002 and 2017.

Narratives from the Illawarra show how in the absence of fathers, who may have died or been left behind in countries of origin, it is often the mothers who bear responsibility for undertaking the long journey to resettlement with their children (Lakhina, 2019). Also, women from refugee backgrounds often under-take the work of intergenerational caring, by creating rituals of remembering and healing across places. In undertaking this kind of analysis, we seek to go beyond the task of situating differential vulnerabilities to climate-related hazards to also understand transformative practices of safe refuge. Our study shows how women from refugee backgrounds rely on their past experiences, traditional knowledge, and everyday practices of care, for self and community, to feel safe and secure as they resettle in Australia. In the final section, we assess the value of adopting intersectional analysis to better understand both the impacts of climate change and opportunities for adaptation, with a view to transform social relations for safe refuge in the 21st century. We argue that understanding how women from diverse refugee backgrounds find safety in unfamiliar places can be instructive for reim-agining climate resilience in transformative and caring ways, for all of humanity.

## Who Belongs? Australian Discourses on Climate Resilience

Australia is rich in diverse cultural and environmental practices. According to the latest census, about 30 per cent of Australia's population was born overseas (Australian Bureau of Statistics [ABS], 2020). However, this diversity is not fully reflected in policy constructions of how Australians understand nature, land-scapes, and climate variability and change. While there has been a slow and long overdue recognition of Indigenous environmental, land, and fire management practices in recent years (Eriksen & Hankins, 2014; Gammage, 2011; Steffensen, 2020), an understanding of diverse migrant environmental practices still remains scarce (Klocker & Head, 2013; Klocker et al., 2018; van Holstein & Head, 2018). For example, the Commonwealth of Australia's *Sustainable Population Strategy* (2011) highlighted the role of Indigenous knowledge in environmental and plan-ning policy but overlooked the diverse knowledge and practices of migrants (Planning Institute of Australia, 2011). More recently, the Commonwealth of

Australia's *National Climate Resilience and Adaptation Strategy* (2015) acknowledges that people facing socio-economic disadvantages, including migrants and refugees, are more vulnerable to all kinds of hazards in Australia. Yet, the strategy does not acknowledge pathways on *how to* engage with people's diverse experiences and capacities to cope and adapt to climate change.

This is, in part, because Australia's environmental debate has been formulated less in the context of understanding people's differential experiences of coping with increasing climate impacts, and more as a policy conundrum on how to accommodate a growing migrant population (Betts, 2004; Klocker & Head, 2013; Missingham et al., 2006). At least since 1996, the Australian government has encouraged newly arrived migrants to settle in rural areas with the dual objective of easing population stress on urban services and infrastructure, while revitalizing rural economies (Missingham et al., 2006). Here, "rural" broadly refers to towns with a population of less than 100,000 people (ABS, 2020). While Australia's regional resettlement policy has, in part, contributed to the economic development of regional economies, it has also led to people from refugee backgrounds becoming invisible in planning agendas (Fozdar & Hartley, 2013). As we show below, such agendas continue to have a lopsided focus on the planning and sustainability needs of metropolitan areas. In 2003, the release of the *Review of Settlement Services for Migrants and Humanitarian Entrants* (Department of Multicultural and Indigenous Affairs, 2003) reiterated the opportunity to encourage refugee resettlement in regional and rural Australia. Of the estimated total number of people who migrated to Australia between 2006 and 2016, including cases of humanitarian resettlement, about 32 per cent settled in NSW, of which about 90 per cent settled in the Greater Sydney metropolitan area (ABS, 2017). The remaining balance, an estimated 76,358 people, settled across NSW in regional and rural areas during this decadal period (ABS, 2017).

What has the regional resettlement policy meant for how people from refugee and migrant backgrounds find safety from climate-related hazards in regional and rural contexts across Australia? Also, to what extent do current planning agendas recognize that people from refugee and migrant backgrounds, respectively, have quite different resettlement experiences in regional Australia due to issues related to language, education, cultural values, and differential access to housing and health services (Fozdar & Hartley, 2013)? Reflecting the national policy context, broadly two kinds of research have been pursued to understand ethnically diverse attitudes and practices with regard to Australia's changing environment and climate. The majority of research with ethnic minorities continues to focus on understanding the planning and sustainability needs of metropolitan areas, mainly Brisbane, Sydney, and Melbourne (Fincher et al., 2014; Head & Gibson, 2012; Osborne, 2015). Within this planning research agenda, Australia's ethnic minorities' environmental concerns and values have been assessed by a small number of urban-centred sustainability indicators, such as household practices of recycling, green consumption attitudes and perceptions, participation in environmental groups, and outdoor recreation (Larson et al., 2011; Maller, 2011). For example, Klocker and Head (2013) examined the ways ethnic diversity has been represented in household scales of analysis around sustainable lifestyles in the context of climate change. More recently, Waitt (2018) and Waitt and Nowroozipour (2018) examined how ethnically diverse populations in metropolitan areas understand their relationship with water scarcity and conservation practices in Sydney.

Many studies on migrant households' environmental practices overwhelmingly rely on western ontologies and settler norms (van Holstein & Head, 2018) to examine evidence of pro-environmental behaviour and forms of environmental activism among ethnic minorities (Larson et al., 2011). For example, the *Moving landscapes* series (Thomas, 2001, 2002) documented the distinctive view of nature that Macedonian and Vietnamese people brought to Australia. The series investigates the Australian idea of environmental management and conservation in the overall context of migrant cultures and journeys, exploring how migrant communities see themselves reflected in the ideal of national parks, and environmental conservation more generally. However, as Klocker and Head (2013) pointed out, any perceptions of disengaged ethnic minorities may reflect language barriers and cultural or religious forms of exclusion in mainstream environmental discourses of, for example, conservation and national parks, rather than a disregard or lack of concern for environmental issues among ethnic minorities.

An emerging body of cultural and environmental research has developed a better understanding of vernacular practices among ethnic minorities in regional and rural Australia (Head et al., 2019; Klocker & Head, 2013; van Holstein & Head, 2018). For example, Klocker et al. (2018) examined the role of immigrant groups in the development of horticultural and gardening practices as they settle in suburban and regional parts of Australia. Such case studies show how ethnic minority farmers in Australia's food bowl—the Murray-Darling basin—have successfully engaged with experimental food growing practices in highly uncertain times. Specifically, they highlight how diverse food growing cultures in post-migration contexts represent a potential adaptive resource, which should be better recognized as Australia adapts to a changing landscape in a new climate. Head et al. (2019) also brought attention to migrant practices in the semi-arid Sunraysia region where, "without effective climate change mitigation, there might only be another thirty seasons of irrigated agriculture possible" (Head et al., 2019, p. 1904). This evolving body of research on diverse food growing cultures focuses on understanding the culturally diverse capacities and resources that migrants bring to Australia's rural landscapes. It provides important empirical insights on how different ethnic groups have different "ways of seeing the Australian natural world" (Head, 2000, pp. 236–237), and "how different groups of people understand broader human-environment relations" in agrarian contexts (Head et al., 2019, p. 1907).

In the next section, we build on these insights to delve into how women from diverse refugee backgrounds cope with climate-related hazards as they resettle in unsafe housing, unfamiliar cultures, and changing landscapes in a particular area of regional Australia. In doing so, we move past the migrant productivity and resourcefulness debates to examine women's beliefs, experiences, and everyday practices of finding safe refuge in Australia. By foregrounding the lived experiences of women from refugee backgrounds, we seek to address a historical lack of understanding of the lived experiences of single-parent families, often headed by women, in times of crises (Enarson & Fordham, 2000). Building on a novel person-centred mapping methodology (Lakhina, 2018, 2019), we glean examples of how women's lived experiences and everyday practices can be an important determinant of their ability to access reliable hazard and risk information, safe housing, and critical emergency services. By revealing women's visceral experiences of structural inequality, systemic discrimination, and lack of access to basic amenities and services, we foreground lived experiences and everyday practices of finding safety and wellbeing across changing landscapes.

## Whose Experiences Count? Narratives of Refuge and Practices of Safety in the Illawarra

The Illawarra is a coastal region south of Sydney comprising three local government areas: Wollongong, Shellharbour, and Kiama. Mirroring Australia's overall migratory trends, the Illawarra has been shaped by four waves of migration since the 1950s (Migration Heritage Centre, 2015). People arrived in the Illawarra from across Europe after the Second World War, during the 1970s from Lebanon, from the Indo-China region and South America in the 1980s. Since the late 1990s, people have mainly been resettled from the Middle East and North Africa, the Horn of Africa and West Africa, South Asia and South-East Asia (Fozdar & Hartley, 2013; Wollongong City Council [WCC], 2020). Of significance, the Illawarra regularly experiences a range of climate-related hazards, including bushfires, heatwaves, storms, flash flooding, and lightning (NSW Government, 2010). Over the coming decades, mean temperatures, rainfall, and sea levels, among other climate impacts, are projected to increase for the Illawarra (NSW Government, 2010, 2018).

As newly arrived people resettle in regional and rural areas across the Illawarra, it is important to develop a better understanding of how the remoteness from emergency and daily settlement support services could affect how they cope with unfolding climate impacts (Lakhina, 2019). For example, rental accommodation offered to newly arrived people from refugee backgrounds in the Illawarra are located in neighbourhoods that rank among NSW's poorest (DOTE, 2015; WCC, 2013, 2020). These neighbourhoods are located in suburbs that have been worst affected by heavy rain, storms, and flash floods in recent years (Pearson, 2017). However, newly arrived people's experiences with poor-quality housing, neighbourhood violence, and exposure to climate impacts remain largely unknown and undocumented in the Illawarra and other regions of resettlement across Australia.

In this section, we present a person-centred approach to examine how women from diverse refugee backgrounds rely on their past experiences, personal and cultural beliefs, and everyday practices to feel safe and secure as they resettle in new housing, landscapes, and cultures. We briefly examine these three kinds of narratives to show how women from refugee backgrounds can experience variegated geographies of climate change in regional Australia. The first narrative reveals a lived experience of *unsafe refuge*, showing how women from refugee backgrounds can experience systemic racism and housing discrimination, as they navigate long spatial and temporal journeys of resettlement moving between coastal, suburban, rural, and inner-city housing. The second narrative emphasizes the role of *communities of healing* to show how women from refugee backgrounds engage in everyday practices of healing and care—for self, community, and distant strangers. The third narrative foregrounds everyday practices that bridge past experiences and traditional forms of knowledge with emerging vernacular practices in uncertain *social geographies of climate resilience*. In the order in which we present the narratives, the research participants migrated from Iran, Liberia, and Burma, respectively. They resettled in the Illawarra between 2002 and 2017.

The three kinds of narratives presented here draw on semi-structured interviews conducted by the lead author between July and September 2017 in the Illawarra. Following approval from the University of Wollongong's Human Research Ethics Committee, a total of 26 semi-structured interviews were conducted with people from diverse refugee backgrounds. The interviews followed three broad themes: 1) moving, settling, and living in the Illawarra; 2) beliefs, attitudes, and experiences with natural hazards, climate, and the environment; and 3) everyday practices for feeling safe and secure. The interviews were conducted in participants' homes, facilitated by community liaisons who provided simultaneous interpretations. The interviews lasted between 45 and 130 minutes and were audio recorded and transcribed verbatim for analysis. The following quotes include a range of English language competencies due to the varying levels of interpretation support provided for each interview. However, as is evident, the different levels of English language do not take away from research participants' ability to narrate the full extent of their experiences of seeking safe refuge in Australia. As a mark of respect for the deeply personal narratives shared by research participants, we use pseudonyms rather than impersonal codes (Lee & Hume-Pratuch, 2013).

### Unsafe Refuge

Zoya arrived in the Illawarra in 2016. A widow from Iran, she came to Australia with her two teenage sons. Her earliest memory is of negotiating with the settlement services for a safe and healthy home. The first house she was shown on arrival was severely affected by damp and mould. She recalled telling the settlement caseworker:

> One of my sons has asthma, I can't live here […] my children are unsafe here.

After a health risk assessment was conducted by the settlement services, Zoya's family was offered a tourist beach cabin for a stipulated period of four weeks. After overstaying almost two months in this temporary accommodation, Zoya finally found a place to rent in a low-income suburb in the Illawarra. She recalled the caseworker warning her, "it was not a good area for living," but she did not feel like she had a choice. She could not find an affordable place in any other neighbourhood at that point. Zoya shared:

[…] normally people who are living there are people who have mental illness, people who can't afford it and people who are drug and alcohol users. The housing department, they put them [there].

She shared how for the next few months, her family suffered racial abuse, violence, and theft. They felt "imprisoned in the house" due to racist threats and neighbourhood violence:

My little son, he developed depression and for three months he didn't want to go out. He developed, I think, post traumatic or something like that. He was just screaming all the night and he didn't want to go to any psychologist or any doctor.

With help from some Iranian friends, Zoya and her sons soon moved again, to a house near the beach. At the time of the interview, they felt safer from neighbourhood violence due to the remoteness of their new accommodation. However, she remained worried about her exposure to the elements:

[It] is a timber house... if that tree that is near the window falls into the window [...] would die. It's not safe house. I think because it's old... It's really, really old... because of a wind it just shakes like this. So, what happens if a tree falls down?... I have no idea what to do after... if there is a fire. So, what do you have to do?

### Communities of Healing

Teta looked for a church almost as soon as she arrived in the Illawarra in 2002:

I started to ask people 'where is my church' [the denomination she attended in Liberia] and I go to my church … I say, 'Oh, I have family here… I have the church here, I have family'. Coming here … it was something that God plans for me. I had no hope of us leaving Africa and … as a refugee … it was lot of struggles and things … So, my coming here is like, 'Oh God, I am going in a new place, like, I know no one. It's only you I depend on'. I know that he would carry me through, so, it was challenging, but… I make it to the end.

Teta's church prayers regularly include strangers and distant others. She described an instance when the women's group in her church came together to pray for people affected by typhoons in the Philippines:

Africans, even though we don't have experience for those things [typhoons], but we always pray, that it shouldn't happen, even though it's … those people it happened to, that they will be able to be safe, Additionally, even though it's not happening to us, it's happening to people … That they should be able to be safe. So … we always pray for that, yeah. Pray that God will continue to protect us. I pray with my family, every day and night, yes. We pray for them … Because it's God that brought us into this nation, so we don't want those thing[s] to come and make us unhappy, so, we always pray that God will continue to protect us.

Teta's church group not only prayed for people affected by the typhoons in the Philippines, but they also attempted to learn from the disaster. They came together to plan for their own preparedness, by talking about the kinds of items they could stock up on, including candles and food. Teta remembers preparing in similar ways for seasonal hazards, like flooding, with her church community back in Liberia.

### Social Geographies of Climate Resilience

Daw arrived with her family of five from Burma in 2012. She relied on gardening as a way of making home in the Illawarra:

Having a garden means power of life, and here buying vegetables is a strange concept to us, because we're used to planting in the forest or in the garden.

Over the years, Daw has adapted her traditional knowledge from Burma to successfully farm in the Illawarra's suburban landscape:

[…] when it rains at night you have to water the next day, otherwise we have insects. Water is also part of cleaning […] or getting rid of insects […] Without their knowledge [ancestors], I will rely on the chemical products to get [rice] without insects, and it affects us indirectly. By knowing that, the next day I just water around the plants and get rid of insects naturally […] in a way it's a bit strange for us, that people use insecticides to kill the insects. I have never done that in my life. I've never used insecticide.

Gardening also becomes a way to bridge ties with newly arrived members of her Burmese community:

We share. We try to share with our friends […] Now we have new arrival refugees, our friends will bring them to the garden to help us […] plant and we share the produce.

However, Daw has struggled to encourage her children to adopt her traditional knowledge of planting. Unlike her agrarian upbringing in Burma, Daw's children were born in a Thai refugee camp. She explained how her children had no direct knowledge of farming until they arrived in Australia and witnessed her home gardening practices:

When they see the soil, they say it's yucky. It's a different generation now… You ask them to pick vegetables or fruits, they don't know when to begin, or where to begin, or what to pick. When you water, there is a certain way of on watering plants. They don't know how to do, so we have lost the lesson. They will be the lost generation.

Yet, she acknowledged her children's help in giving her daily weather reports and timely forecasts from weather apps on their mobile phones. Daw does not read or write, so daily acts of intergenerational knowledge sharing helped her garden survive even during uncertain weather, winds, and rain in past years.

The narratives presented above reveal how women from refugee backgrounds rely on their experiences, traditional knowledge, and everyday practices of care— for self and community, to feel safe and secure as they resettle in Australia. This highlights how an intersectional analysis of climate-related hazards requires us to not just ask who is affected by climate change impacts, but also to engage with empowering and transformative narratives that reveal how people cope and adapt across changing landscapes. A sense of community is forged daily for people from diverse refugee backgrounds. There is no homogenous or one well-organised community of belonging, rather identities can be fractured, tangential, and context specific, also extending to distant strangers. Teta's narrative, for example, emphasizes how women from refugee backgrounds engage in everyday practices of care—for self, community, and distant strangers. The narratives also show us that diverse beliefs in "sacred ecologies" (Northcott, 2015) help women from refugee backgrounds find belonging and safety in unfamiliar landscapes. Daw's narrative shows how cultural attitudes to landscapes can be dynamic across generations and spatial contexts. These narratives reveal how cultural narratives, personal beliefs, and everyday practices are often developed in response to a range of ongoing social, economic, political, and climatic changes. Adopting an intersectional lens has enabled us to go beyond examining social vulnerabilities to also recognize the transformative aspects of social resilience. Understanding these kinds of diverse experiences can inform the development of socially just and caring forms of disaster research, policy, and practice. The transformative pathways transform social relations through a mutual process of co-learning, which entails systematically informing, engaging, and partnering with people from diverse refugee backgrounds for disaster resilience (Lakhina, 2019).

The three kinds of narratives presented above also show how newly arrived people can face unequal exposure to climate-related hazards due to structural inequalities that result in or exacerbate unsafe housing, unfamiliar cultures, and changing landscapes. These experiences can reinforce conditions that exclude people from refugee and migrant backgrounds from disaster resilience and climate adaptation policy and actions. People from diverse refugee backgrounds shared experiences of finding unsafe refuge in Australia. Ten of the 26 research participants were caught unaware by bushfires, flooding, hail, rain, lightning, and strong coastal winds within their first year of arriving in the Illawarra. Nine of the

26 research participants lived in what they perceived to be unsafe, insecure, and poorly maintained housing during their initial months (Lakhina, 2018). Some participants, such as Zoya, felt traumatized by unwelcoming neighbourhoods, where they experienced rampant drug abuse, alcoholism, and related violence. These experiences highlight the need for providing culturally appropriate and timely risk information as well as community and support services for socially and physically isolated households, and for creating opportunities for newly arrived people to train with the local emergency services.

When these research findings were initially released in 2017, it created opportunities for community representatives from diverse refugee backgrounds to discuss needs and next steps with local institutional representatives from the city council, emergency services, and multicultural services. This resulted in the creation of the first-ever Multicultural Liaison Unit in the NSW State Emergency Services (SES)—Illawarra region. The unit, formed in December 2017 as part of the NSW SES Volunteering Reimagined programme, is comprised of multicultural community liaison officers from diverse refugee backgrounds trained to communicate about natural hazards and local safety procedures to other community members from refugee backgrounds (Lakhina et al., 2019). As of 2021, the Unit continues to work with volunteers from diverse refugee backgrounds to provide culturally and linguistically appropriate preparedness and outreach services to emerging communities across the Illawarra. As part of Australia's Harmony Week celebrations (15–21 March 2021), a volunteer shared her experience:

> If I was not part of NSW SES, I don't think my community would know what to do in a flood or storm. I'm not sure that NSW SES previously reached out to people of Karenni, Karen, or Burmese backgrounds to interact in a culturally sensitive way. Through the Multicultural Community Liaison Unit, people from the community feel more empowered—they feel they have a voice and can bring their experiences to the attention of NSW SES. (NSW SES, 2021)

This research outcome demonstrates concrete pathways to ground disaster policy and programmes in a better understanding of how people, particularly women and mothers, from diverse refugee backgrounds practice safety in unfamiliar places. Developing this kind of intersectional understanding of gender, culture, and immigration status, can be instructive for reimagining climate resilience in transformative and caring ways for all of humanity.

Narratives from the Illawarra also prompt further questions for future work. How do newly arrived people from refugee backgrounds practice food security in regional and rural areas during droughts, flooding, or destructive winds? How do culturally diverse communities engage with changing weather patterns and uncertain environmental cycles? How do diverse forms of climate mobilities enable people to cope and adapt across (changing) landscapes? By design, such research questions should reflect on people's experiences with a wide range of mobilities, across borders, landscapes, and housing.

## Conclusion

Following recent attempts to outline an intersectional framework for climate-related hazards research (Walker et al., 2019), this chapter pursued three broad objectives for an intersectional analysis of climate resilience in the Illawarra. We have: 1) centred people's lived experiences, beliefs, and everyday practices, 2) explored pathways to ground disaster policy in people's everyday lived experiences, and 3) engaged with our research in caring ways (Lakhina, 2019). Through illuminating quotes from research participants' interview narratives, we presented empirical evidence for how an intersectional analysis can engage with gender, culture, and immigration status to reveal important insights for grounding research, policy, and practice in people's lived experiences and everyday practices of safety. These lived experiences and everyday practices often determine people's ability to access reliable hazard and risk information, safe housing, and critical emergency services. Research participants' narratives show how people from diverse refugee backgrounds engage with changing landscapes in contingent ways, relying on their experiences, beliefs, and cultural knowledge across generations. These insights afforded regional authorities and local emergency services with new entry points for systematically informing, engaging, and partnering with people from diverse refugee and migrant backgrounds in the Illawarra (Lakhina et al., 2019). We also demonstrate how attending to intersectional analysis can give voice to women from diverse refugee and migrant backgrounds, who otherwise remain invisible in current representations of the changing Australian climate and landscape (Klocker & Head, 2013; van Holstein & Head, 2018).

Looking to an uncertain future, our findings are about more than how people from refugee backgrounds experience safety in unfamiliar and changing landscapes. In this "age of migration" (Castles & Miller, 2009), adopting an intersectional lens can reveal how certain identities and mobilities are not only "stigmatized, trivialized, valued, or recognized in relation to others" (Fincher et al., 2014, p. 3), but can be critical to developing a caring approach to climate resilience. In attending to multiple scales of intersectional analysis, we conclude by suggesting that an inclusive climate resilience agenda can be partly addressed by developing sustained pathways to systematically inform, engage, and partner with people from refugee and migrant backgrounds. For future intersectional analysis centred on climate-related hazards, we recommend exploring pathways to co-learning disaster resilience with people from diverse refugee and migrant backgrounds (Lakhina, 2019). Finally, engaging with the social geographies of climate resilience will require extending a current focus on urban planning processes to include regional, rural, and agrarian contexts, not as binaries, but as contiguous sites of climate mobilities. It will require an understanding of how climate mobilities encompass a diversity of cultural and environmental practices from the inner city and suburban gardens, to farming practices in rural regions across the Global North. It is between these liminal sites of refuge that emotive and everyday relationalities with a changing climate can be located.

# References

Agarwal, B. (1992). The gender and environment debate: Lessons from India. *Feminist Studies*, *18*(1), 119–158. https://doi.org/10.2307/3178217

Alston, M. (2011). Gender and climate change in Australia. *Journal of Sociology*, *47*(1), 53–70. https://doi.org/10.1177/1440783310376848

Australian Bureau of Statistics. (2017, July 20). *Census of population and housing reflecting Australia. Stories from the Census 2016 – cultural diversity [Table 11. Year of arrival in Australia by Greater capital City Statistical areas, count of persons]*. https://www.abs.gov.au/AUSSTATS/abs@.nsf/DetailsPage/3415.02020?OpenDocument

Australian Bureau of Statistics. (2020, June 11). *Australian Statistical Geography Standard*. https://www.abs.gov.au/websitedbs/D3310114.nsf/home/Australian+Statistical+Geography+Standard+(ASGS)

Baldwin, J. (2017). Decolonising geographical knowledges: The incommensurable, the university and democracy. *Area*, *49*(3), 329–331. https://doi.org/10.1111/area.12374

Betts, K. (2004). Demographic and social research on the population and environment nexus in Australia: Explaining the gap. *Population and Environment*, *26*(2), pp. 157–172. https://doi.org/10.1007/s11111-004-0838-9

Boas, I., Farbotko, C., & Adams, H. (2019). Climate migration myths. *Nature Climate Change*, *9*, 901–903. https://doi.org/10.1038/s41558-019-0633-3

Castles, S., & Miller, M. J. (2009). *The age of migration: International population movements in the modern world* (4th ed.). Palgrave MacMillan.

Cho, S., Crenshaw, K., & McCall, L. (2013). Toward a field of intersectionality studies: Theory, applications, and praxis. *Signs*, *38*(4), 785–810. https://doi.org/10.1086/669608

Clement, V. (2019). Beyond the sham of the emancipatory enlightenment: Rethinking the relationship of Indigenous epistemologies, knowledges, and geography through decolonizing paths. *Progress in Human Geography*, *43*(2), 276–294. https://doi.org/10.1177/0309132517747315

Collins, P. H. (1990). *Black feminist thought: Knowledge, consciousness and the politics of empowerment*. Unwin Hyman.

Collins, P. H. (2015). Intersectionality's definitional dilemmas. *Annual Review of Sociology*, *41*, 1–20. https://doi.org/10.1146/annurev-soc-073014-112142

Collins, P. H., & Bilge, S. (2020). *Intersectionality* (2nd ed.). Polity, Wiley.

Commonwealth of Australia. (2011). *Sustainable Australia: Sustainable communities: A sustainable population strategy for Australia*. Department of Sustainability, Environment, Water, Population and Communities, Canberra. https://apo.org.au/sites/default/files/resource-files/2011-03/apo-nid166281.pdf

Commonwealth of Australia. (2015). *National climate resilience and adaptation strategy*. Department of the Environment, Canberra. https://www.environment.gov.au/system/files/resources/3b44e21e-2a78-4809-87c7-a1386e350c29/files/national-climate-resilience-and-adaptation-strategy.pdf

Connell, R. (2007). *Southern theory: The global dynamics of knowledge in social science*. Polity Press.

Crenshaw, K. W. (1991). Mapping the margins: Intersectionality, identity politics and violence against women of color. *Stanford Law Review*, *43*(6), 1241–1299. https://doi.org/10.2307/1229039

Das, V. (1995). *Critical events: An anthropological perspective on contemporary India*. Oxford University Press.

Department of Multicultural and Indigenous Affairs. (2003). *Report of the review of settlement services for migrants and humanitarian entrants.* Canberra.

Djoudi, H., Locatelli, B., Vaast, C., Asher, K., Brockhaus, M., & Basnett Sijapati, B. (2016). Beyond dichotomies: Gender and intersecting inequalities in climate change studies. *Ambio, 45*(S3), 248–262. https://doi.org/10.1007/s13280-016-0825-2

DOTE. (2015). *Dropping off the edge: Persistent communal disadvantage in Australia.* Jesuit Social Services and Catholic Social Services Australia. https://dote.org.au/findings/full-report/

Enarson, E. (1998). Through women's eyes: A gendered research agenda for disaster social science. *Disasters, 22*(2), 157–173. https://doi.org/10.1111/1467-7717.00083

Enarson, E., & Fordham, M. (2000). Lines that divide, ties that bind: Race, class, and gender in women's flood recovery in the US and UK. *Australian Journal of Emergency Management, 15*(4), 43–52.

Eriksen, C. (2014). *Gender and wildfire: Landscapes of uncertainty.* Routledge.

Eriksen, C., & Hankins, D. L. (2014). The retention, revival and subjugation of Indigenous fire knowledge through agency fire fighting in eastern Australia and California, USA. *Society & Natural Resources, 27*, 1288–1303. https://doi.org/10.1080/08941920.2014.918226

Fiddian-Qasmiyeh, E. (2020). *Refuge in a moving world: Tracing refugee and migrant journeys across disciplines.* UCL Press.

Fincher, R., Iveson, K., Leitner, H., & Preston, V. (2014). Planning in the multicultural city: Celebrating diversity or reinforcing difference? *Progress in Planning, 92*, 1–55. https://doi.org/10.1016/j.progress.2013.04.001

Fozdar, F., & Hartley, L. (2013). Refugee resettlement in Australia: What we know and what we need to know. *Refugee Survey Quarterly, 32*(3), 23–51. https://doi.org/10.1093/rsq/hdt009

Gammage, B. (2011). *The biggest estate on Earth: How Aborigines made Australia.* Allen & Unwin.

Giddens, A. (2009). *The politics of climate change.* Polity Press.

Gill, N. (2018). The suppression of welcome. *Fennia, 196*(1), 88–98. https://doi.org/10.11143/fennia.70040

Guha, R. (Ed.). (1982). *Subaltern studies I: Writings on South Asian history and society.* Oxford University Press.

Gumbs, A., Martens, C., & Williams, M. (2016). *Revolutionary mothering: Love on the frontlines.* Between the Lines.

Harris, M. L. (2016). Ecowomanism: An introduction. *Worldviews, Global Religion, Culture and Ecology, 20*(1), 5–14.

Head, L. (2000). *Second nature: The history and implications of Australia as Aboriginal landscape.* Syracuse University Press.

Head, L., Klocker, N., Dun, O., & Aguirre-Bielschowsky, I. (2019). Cultivating engagements: Ethnic minority migrants, agriculture, and environment in the Murray-Darling Basin, Australia. *Annals of the American Association of Geographers, 109*(6), 1903–1921. https://doi.org/10.1080/24694452.2019.1587286

Head, L. M., & Gibson, C. R. (2012). Becoming differently modern: Geographic contributions to a generative climate politics. *Progress in Human Geography, 36*(6), 699–714. https://doi.org/10.1177/0309132512438162

hooks, b. (1990). *Yearning: Race, gender and cultural politics.* South End Press.

hooks, b. (2000). *Feminism is for everybody: Passionate politics.* Pluto Press.

Jacobs, F. (2019). Black feminism and radical planning: New directions for disaster planning research. *Planning Theory, 18*(1), 24–39. https://doi.org/10.1177/1473095218763221

Kaijser, A., & Kronsell, A. (2014). Climate change through the lens of intersectionality. *Environmental Politics, 23*(3), 417–433. https://doi.org/10.1080/09644016.2013.835203

Kelman, I. (2020). Does climate change cause migration? In E. Fiddian-Qasmiyeh (Ed.), *Refuge in a moving world: Tracing refugee and migrant journeys across disciplines.* UCL Press.

Klocker, N., & Head, L. (2013). Diversifying ethnicity in Australia's population and environment debates. *Australian Geographer, 44*(1), 41–62. https://doi.org/10.1080/00049182.2013.765347

Klocker, N., Head, L., Dun, O., & Spaven, T. (2018). Experimenting with agricultural diversity: Migrant knowledge as a resource for climate change adaptation. *Journal of Rural Studies, 57*, 13–24. https://doi.org/10.1016/j.jrurstud.2017.10.006

Lakhina, S. J. (2018). *Co-learning disaster resilience toolkit: A person-centred approach to engaging with refugee narratives and practices of safety.* University of Wollongong.

Lakhina, S. J. (2019). Co-learning disaster resilience: A person-centred approach to understanding experiences of refuge and practices of safety [Doctor of Philosophy thesis, School of Geography and Sustainable Communities, University of Wollongong. https://ro.uow.edu.au/theses1/860

Lakhina, S. J., Eriksen, C., Thompson, J., Aldunate, R., McLaren, J., & Reddy, S. (2019). People from refugee backgrounds contribute to a disaster-resilient Illawarra. *Australian Journal of Emergency Management, 34*(2), 19–20.

Lama, P., Hamza, M., & Wester, M. (2020). Gendered dimensions of migration in relation to climate change. *Climate and Development, 13*(4), 326–336. https://doi.org/10.1080/17565529.2020.1772708

Larson, L., Whiting, J., & Green, G. (2011). Exploring the influence of outdoor recreation participation on pro-environmental behaviour in a demographically diverse sample. *Local Environment, 16*(1), 67–86. https://doi.org/10.1080/13549839.2010.548373

Lee, C., & Hume-Pratuch, J. (2013, August 22). Let's talk about research participants. APA Style Blog (6th ed.) Archive. http://blog.apastyle.org/apastyle/2013/08/lets-talk-about-research-participants.html

MacGregor, S. (2006). *Beyond mothering Earth.* UBC Press.

MacGregor, S. (2010). A stranger silence still: The need for feminist social research on climate change. *Sociological Review, 57*, 124–140. https://doi.org/10.1111/j.1467-954X.2010.01889.x

Maller, C. (2011). Practices involving energy and water consumption in migrant households. In P. W. Newton (Ed.), *Urban consumption* (pp. 237–249). CSIRO Publishing.

Migration Heritage Centre. (2015). *Every story counts! Illawarra migration heritage project.* New South Wales, Australia. www.migrationheritage.nsw.gov.au/projects/current-project/investigating-the- migration-heritage-of-wollongong/

Missingham, B., Dibden, J., & Cocklin, C. (2006). A multicultural countryside? Ethnic minorities in rural Australia. *Rural Society, 16*(2), 131–150. https://doi.org/10.5172/rsj.351.16.2.131

Mohanty, C. T. (1991). Under western eyes: Feminist scholarship and colonial discourses. *Boundary 2, 12*(3), 333–358. https://doi.org/10.2307/302821

Moraga, C., & Anzaldua, G. (Eds.). (1981). *This bridge called my back: Writings by radical women of color.* Persephone Press.

New South Wales Government. (2010). *Impacts of climate change on natural hazards profile: Illawarra region.* NSW Department of Environment, Climate Change and Water, Wollongong.

New South Wales Government. (2018). *Illawarra climate change snapshot.* New South Wales Department of Environment, Climate Change and Water, Wollongong. http://climatechange.environment.nsw.gov.au/Climate-projections-for-NSW/Climate-projections-for-your-region/Illawarra-Climate-Change-Downloads

New South Wales State Emergency Service. (2021). *Hall of fame.* Multicultural Community Liaison Unit. https://www.ses.nsw.gov.au/get-involved/volunteer/volunteers-list/elizabeth-jowanie/

Northcott, M. S. (2015). *Place, ecology and the sacred: The moral geography of sustainable communities.* Bloomsbury Publishing.

Osborne, N. (2015). Intersectionality and kyriarchy: A framework for approaching power and social justice in planning and climate change adaptation. *Planning Theory, 14*(2), 130–151. https://doi.org/10.1177/1473095213516443

Pearson, A. (2017). *The Illawarra's most storm-affected suburbs revealed.* Illawarra Mercury. https://www.illawarramercury.com.au/story/5012284/the-illawarras-most-storm-affected-suburbs-revealed/

Planning Institute of Australia. (2011). *A sustainable population for Australia: Issues paper.* https://www.planning.org.au/documents/item/2591

Quarantelli, E. L. (1992). *The importance of thinking of disasters as social phenomena* (Preliminary Report No. 184). University of Delaware Disaster Research Center. http://udspace.udel.edu/handle/19716/572

Reed, M. G., Scott, A., Natcher, D., & Johnston, M. (2014). Linking gender, climate change, adaptive capacity, and forest-based communities in Canada. *Canadian Journal of Forest Resources, 44*(9), 995–1004. https://doi.org/10.1139/cjfr-2014-0174

Ryder, S. S. (2017). A bridge to challenging environmental inequality: Intersectionality, environmental justice, and disaster vulnerability. *Social Thought and Research, 34*, 85–115. https://doi.org/10.17161/1808.25571

Shiva, V. (1989). *Staying alive: Women, ecology and development.* Zed Books.

Spivak, G. C. (1988). Can the subaltern speak? In C. Nelson & L. Grossberg (Eds.), *Marxism and the interpretation of culture.* Macmillan.

Steffensen, V. (2020). *Fire country: How Indigenous fire management could help save Australia.* Hardie Grant Travel.

Thomas, M. (2001). *A multicultural landscape: National parks and the Macedonian experience.* New South Wales National Parks and Wildlife Service. Pluto Press.

Thomas, M. (2002). *Moving landscapes: National parks and the Vietnamese experience.* New South Wales National Parks and Wildlife Service. Pluto Press.

United Nations. (1994, May 23–27). Yokohama strategy and plan of action for a safer world. Guidelines for natural disaster prevention, preparedness and mitigation. In [Conference session]. World Conference on Natural Disaster Reduction, Yokohama, Japan.

United Nations. (2005). *Hyogo framework for action 2005–2015: Building the resilience of nations and communities to disasters.* https://www.unisdr.org/2005/wcdr/intergover/official-doc/L-docs/Hyogo-framework-for-action-english.pdf

United Nations. (2015). *Sendai framework for disaster risk reduction (2015–2030).* https://www.preventionweb.net/files/43291_sendaiframeworkfordrren.pdf

van Holstein, E., & Head, L. (2018). Shifting settler-colonial discourses of environmentalism: Representations of indigeneity and migration in Australian conservation. *Geoforum, 94*, 41–52. https://doi.org/10.1016/j.geoforum.2018.06.005

Waitt, G. (2018). Ethnic diversity, scarcity and drinking water: A provocation to rethink provisioning metropolitan mains water. *Australian Geographer*, *49*(2), 273–290. https://doi.org/10.1080/00049182.2017.1394805

Waitt, G., & Nowroozipour, F. S. (2018). Embodied geographies of domesticated water: Transcorporeality, translocality and moral terrains. *Social and Cultural Geography*, *21*(9), 1268–1286. https://doi.org/10.1080/14649365.2018.1550582

Walker, H. M., Culham, A., Fletcher, A. J., & Reed, M. G. (2019). Social dimensions of climate-related hazards in rural communities of the Global North: An intersectionality framework. *Journal of Rural Studies*, *72*, 1–10. https://doi.org/10.1016/j.jrurstud.2019.09.012

Whittaker, J., Eriksen, C., & Haynes, K. (2016). Gendered responses to the 2009 Black Saturday bushfires in Victoria, Australia. *Geographical Research*, *54*(2), 203–215. https://doi.org/10.1111/1745-5871.12162

Williams, M. (2019). Radical mothering as a pathway to liberation. *Millennium Journal of International Studies*, *47*(3), 497–512. https://doi.org/10.1177/0305829819852418

Wisner, B. (1998). Marginality and vulnerability: Why the homeless of Tokyo don't 'count' in disaster preparations. *Applied Geography*, *18*(1), 25–33. https://doi.org/10.1016/S0143-6228(97)00043-X

Wisner, B., Blaikie, P., Cannon, T., & Davis, I. (1994). *At risk: Natural hazards, people's vulnerability and disasters*. Routledge.

Wollongong City Council. (2013). *Wollongong's preparedness to climate change adaptation for Wollongong's vulnerable groups: A briefing paper*. Community Cultural and Economic Development Council, Wollongong.

Wollongong City Council. (2020). *Refugee communities in the Illawarra*. Wollongong City Council, Wollongong. https://www.wollongong.nsw.gov.au/__data/assets/pdf_file/0013/4333/Refugee-Communities-2020.pdf

# Reflection on Chapter 2

## Inclusion at the Intersections: From Individual to International, Intention to Impact

*Sherryl Reddy*

*Sherryl Reddy has worked in international aid and humanitarian protection for over a decade in various countries across Africa, Asia, and the Middle East with non-government organizations, community organizations, and United Nations agencies. Her area of work focused on safety, wellbeing, and dignity for marginalized groups and vulnerable populations in crisis-affected contexts. This includes women, children, older people, people living with disability, Indigenous community members, and people from minority cultural, linguistic, or faith backgrounds. From 2015 to 2019, Sherryl led a volunteer-powered refugee support organization focused on building a sense of belonging for former refugees settling in the Illawarra region of New South Wales, Australia. In 2019, she joined the New South Wales State Emergency Service as Diversity & Inclusion (D&I) Officer, supporting development and integration of D&I into organizational culture, emergency response, and public engagement. In 2021, Sherryl became the D&I Manager at a public Arts & Culture institution in New South Wales. Sherryl is a woman of colour, an Australian of Indian descent born in Fiji, and a migrant to a land that always was and always will be Aboriginal Land.*

## Author note

Reflections on this chapter are my own personal reflections, drawing on a range of professional experiences across government, non-government and inter-governmental sectors in the field of D&I, working with communities and groups who are historically marginalized, under-represented, and at risk of discrimination, isolation, or exclusion based on social inequalities and intersecting aspects of their identity.

DOI: 10.4324/9781003089209-5

## Overview

Lakhina and Eriksen's chapter resonates with key areas of my experience as a practitioner in several fields including (1) international humanitarian protection, where community-based protection strategies are core to effective protection (e.g., Berry & Reddy, 2010); (2) refugee resettlement and community development programmes, where strengths-based approaches to individual, family, and community resilience are widely promoted (see, for example, www.scarfsupport.org.au/); and (3) diversity and inclusion (D&I) efforts in the emergency services sector in the Global North, where increasing attention is placed on community partnerships for resilience-building before, during, and after climate-related extreme events such as floods and bushfires (e.g., Lakhina et al., 2019).

The language, lens, and concept of intersectionality cuts across these fields. Intersectionality focuses on aspects of identity and representation—such as age, disability, gender, language, culture, ethnicity, Indigenous descent—and systems of power and privilege that pervade social relations across industries and sectors in the Global North, as in the Global South.

Lakhina and Eriksen's chapter highlights the limited levels of institutional engagement with, and representation of, the voices and experiences of minority populations in Australia (in particular, women from refugee backgrounds). Their chapter shows how people from refugee and migrant backgrounds embody diverse lived experiences, hold traditional and place-based knowledge of the changing climate, and can contribute significantly to community-centred preparedness and response initiatives in collaboration with emergency service organizations.

The chapter also reveals the differential impacts of climate-related hazards on humanitarian refugee entrants who are resettled across the Global North and often face a lack of accessible, appropriate, and timely information and support as they adapt to unfamiliar environments. At the same time, the analysis demonstrates the capabilities of people from refugee backgrounds to engage in care and protection strategies for themselves, their families, and their communities as part of adapting to social environments, systems, and practices structured by dominant groups for dominant identities.

An intentional proactive focus on increasing participation and representation of people from diverse backgrounds with intersecting marginalized identities is certainly a feature of contemporary practice in international humanitarian protection initiatives, Australian refugee resettlement and community development programmes, and emergency service sector commitments. This can be seen in evolving research, strategies, and statements on diversity and inclusion—across industries in the Global North—with a focus on D&I in the workplace, in business operations, and in public engagement (e.g., New South Wales State Emergency Service, 2020).

This D&I approach, in mindset and in practice, plays out in different ways across the aforementioned fields of work, interacting with international, inter-cultural, and institutional power structures with impacts that may be instructive for emerging climate justice research and practice. The following sections examine inclusion initiatives in three sectors—all of which are affected by, and respond to, climate change.

## International Humanitarian Protection

Evolving practice in the international humanitarian protection arena over the past two decades has seen significant shifts towards person-centred approaches, strengths-based strategies, and resilience-oriented language. Narratives are focused on capability rather than deficits when international aid agencies respond to the needs of refugee and internally displaced populations in situations of conflict or disaster.

Inter-governmental systems and inter-agency fora indicate:

- A political willingness across the international community for improved partnerships, cooperation, equitable responsibility-sharing, and solidarity with displaced persons and refugees (e.g., Global Compact on Refugees); and
- A commitment to directly engaging local actors—community leaders, organizations, and so-called "vulnerable" groups such as women, children, people with disability, and older people—as partners in decision making, design, and delivery of interventions affecting them.

This commitment is most often enlivened by ensuring diverse representation on international panels and committees and strengthening partnerships with local authorities and grass-roots community organizers "on the ground." These individuals and groups are not only the first responders to climate- and conflict-related emergencies, but also the enduring resident responders supporting ongoing recovery and resilience-building long after international aid agencies have flown in and flown out.

Local actors are certainly more vocal and visible "at the table" of international humanitarian policy and programming than in the past, but has this changed power dynamics in the world of humanitarian aid and development? Arguably, the dominant perception of local actors is one that is limited to "local" expertise. The attribute and ascribed role of "local actor" is situated in a descending hierarchy of international, national, and local agency. Local actors are enabled to speak to local issues and represent local communities. However, in wider deliberations on international humanitarian protection, their knowledge, experience, and expertise is marginalized in an environment where power, status, visibility, and airtime of actors is still determined by pervasive and, arguably, paternalistic notions of "international" versus "local"; "us" and "them"; the Global North assisting the Global South; all of which are framed as mutually exclusive lenses of engagement.

This fragmentation of humanitarian actors is further exacerbated by competing agendas and "patch protection" between international NGOs and between local NGOs, where a focus on "who we are" as an agency often outweighs a truly collaborative approach to "what we do" in a way that centres and best meets the needs of the communities we serve.

## Refugee Resettlement and Community Development

In the Australian community context, similar dynamics play out through a different hierarchy of voices and narratives. The dominant "local" voice represents

those of Ango-Celtic origin who make up the majority of the Australian population. And it is this "local" perspective that is systemically prioritized and centred
in social, cultural, political, and economic contexts. The traditional knowledge and
lived experiences of First Nations/Aboriginal and Torres Strait Islander Peoples
as owners and custodians of Country, as well as the "international" knowledge
and lived experiences of refugee and migrant populations, are comparatively
diminished.

In my work with SCARF Refugee Support (www.scarfsupport.org.au/)—a
volunteer community organization in the Illawarra region of New South Wales,
Australia—we worked with local volunteers to deliver a range of programmes to
promote social and economic inclusion for humanitarian refugee entrants from
14 countries of origin across Asia, Africa, and the Middle East. A core principle
underlying our work was to ground our service-delivery and our communications
approach in mutually supportive, empowering partnerships between former refugees
and local volunteers to strengthen wellbeing, dignity, and inclusion for everyone.

This conscious partnership approach sought to counter "safety net" language
and models of paternalistic outreach that were previously pervasive in the char-
ity sector. When engaging former refugees in research or education activities
and cultural events aimed at building wider community awareness of the refugee experience, and promoting the positive contributions of refugee entrants to
the Illawarra region and to Australia, we held fast to a commitment that community members from refugee backgrounds would be paid an honorarium for
their time, expertise, and lived experience. This approach was challenging for
many government and non-government organizations who requested (and often
expected) former refugees to freely share their personal stories out of a sense
of gratitude.

While paying honorariums was a significant step forward in recognizing the
knowledge and lived experience of former refugees, it took a further two years
before we hired community members from refugee backgrounds as salaried
SCARF staff, recognizing their vital work as community organizers and mobilizers, without which SCARF would not have been able to deliver its services and
programs. Engaging former refugees in decision-making leadership positions or
as Board members influencing the direction of the organization, however, remains
a strategic and operational aspiration.

## Emergency Services Sector

In the emergency services sector, an ongoing challenge for Australian emergency
service organizations (ESOs) has been diversifying staff and volunteer bases to
reflect the diverse communities they serve, including Australian residents and citizens from refugee and migrant backgrounds. Membership and leadership has traditionally been male-dominated and Anglo/Euro-centric in composition. In addition,
emergency services have traditionally portrayed their service roles and personnel
as "heroic saviours" rescuing people in crisis—a (mis)perception that is often com-
pounded by ESO interactions that discount the contribution of community members

from diverse and marginalized backgrounds in keeping themselves, their families, and communities safe before, during, and after emergency events.

This representation is gradually shifting with community-based research, targeted recruitment programmes, and the explicit integration of D&I priorities and actions as part of strategic planning, community partnerships, and collaborative resilience-building approaches (e.g., AFAC, 2020; Young et al., 2020).

Research conducted by Lakhina (2019) in the Illawarra region of New South Wales highlighted the need to engage the knowledge and capacities of people from refugee/migrant backgrounds in emergency awareness, preparedness, and response activities for a disaster-resilient Illawarra. In an effort to translate research into practice, SCARF Refugee Support, in collaboration with the NSW State Emergency Service (NSW SES), established a team of multilingual volunteers from diverse refugee backgrounds—a Multicultural Community Liaison Unit (MCLU).

Through the MCLU, the ten founding volunteers sought to utilize and share their language skills, cultural knowledge, and community connection expertise with local authorities, humanitarian settlement agencies, and ESOs. Their aim was to improve engagement and communication with culturally and linguistically diverse (CALD) communities in a partnership approach to preparedness for flood, storm, tsunami, fire, accident/injury, crime, and public safety emergencies. Members of the MCLU envisioned an embedded, mutually empowering, multi-pronged partnership to:

(i)   Build connections between ESOs and diverse communities across NSW;
(ii)  Build disaster awareness and resilience among diverse communities; and
(iii) Strengthen the inclusion capability of ESOs in their preparedness, response, and recovery efforts.

To date, however, there is no clear intra- or inter-institutional framework for the MCLU's operation. Volunteer members remain keen to operationalize the MCLU's potential. Systemic policy and programme support is needed to recognize and respond to this gap in ESO capability and community reach. Institutional commitment is needed to recognize the value of intrinsic knowledge, skills, and community connections in reaching marginalized communities. Furthermore, institutional leadership is needed to invest in strong partnerships to enhance safety, minimize harm, and maximize community empowerment in responding to climate-related emergencies such as floods, droughts, storms, heatwaves, and bushfires—all of which are increasing in duration and intensity in many parts of Australia, and across the Global North (e.g., AFAC, 2018).

A genuine commitment to community resilience-building requires an intersectional lens applied to people, place, and purpose. Volunteer members of the MCLU saw the initiative as a way to build inclusion skills and resources within and across ESOs—by establishing multi-hazard, cross-agency teams reflecting key marginalized or under-represented groups (e.g., women, young people, older people, Aboriginal communities, CALD communities, people with disability, etc.) to support ESOs state-wide with language, culture, and connection capability. Efficiencies of scale, influence, risk management, and community impact could be achieved with "inclusion capability teams" of this nature, supported by a collaborative, cross-agency approach to recruiting, managing, training, learning from, activating, and supporting these specialist personnel.

This approach challenges traditional agency boundaries and "command and control" ways of working in the emergency services sector. It also calls on us to interrogate organizational cultures and question entrenched assumptions about whose knowledge, expertise, and capability should contribute to institutional and societal change.

With the recent creation of Resilience New South Wales—a state authority leading whole-of-New South Wales disaster and emergency efforts from prevention through to recovery—there may be opportunities for initiatives like the MCLU to be realized as critical community mobilizers and institutional advisors, providing "co-badged" information, communication, and liaison support to communities, across all hazards and emergencies, and continually building the intercultural understanding and capability of emergency services state-wide.

## Intersectional Inclusion: Reflections on Intention and Impact

Intersectionality as a concept is most often focused on personal identities, with less attention to systemic power imbalances. Such systemic issues are pervasive in workplaces and the wider community, and are replicated across agencies and authorities at multiple levels. Institutional aspirations to "better reflect the communities we serve" often veil a reluctance to engage in critical self-reflection and pursue equity actions that challenge power dynamics within and across organizational hierarchies and infrastructure.

Reflecting on progress and practice in these areas has challenged me to question how my own work in this space may have unwittingly fed dominant power structures. Researchers and practitioners should question how well-intended efforts to promote intersectional inclusion can also perpetuate systems that dismiss, minimize, or exclude minority voices and lived experiences from strategic discourse and decision making in the Global North.

We need to confront the gaps that exist between intention and impact; between institutional perceptions and lived reality; between *what* we do and *how* we do it. We need to look critically and openly, with curiosity and humility, at systemic inequality in the processes we employ and the lenses we apply to our research and practice in organizational and inter-agency contexts.

Practices that undermine intersectional inclusion and reinforce existing discriminatory systems include:

- Actions that increase surface-level diversity but retain power in traditional hierarchies;
- Institution-centred community engagement activities that focus on consultation, participation, and collaboration with people from under-represented

backgrounds, but deny them equal agency in the development and leadership of initiatives affecting them and the global community;

- Organizational deference to majority "culture-fit" voices, narratives, knowledge, and experiences;
- Reactive, rather than proactive, engagement in transformational practices to centre the voices and lived experience of people from marginalized backgrounds;
- An expectation that people from marginalized backgrounds share their lived experience in a voluntary capacity, without remuneration, diminishing the value of their contributions compared to salaried staff or paid consultants;
- Limited institutional understanding of the impact of identity strain on people from marginalized backgrounds, combined with a lack of visible, vocal allyship from those with power to influence change;
- Engagement approaches that presumptively define the expertise of people from marginalized backgrounds as limited to their under-represented identity or community, and not extending to local or global issues beyond;
- A lack of will to tackle deep structural barriers and create more egalitarian systems and institutional cultures with increased representation of people from under-represented backgrounds at all levels.

Intersectional inclusion requires structural and behavioural change in which people from diverse and under-represented backgrounds are included as co-creators, mobilizers, and leaders of individual and collective social change—whether that's in a refugee camp setting; in community engagement programmes to prepare for, and respond to, floods and fires in a local region; through research that catalyzes dialogue with new voices; or in the narratives we create and communicate through, and to, government authorities and public institutions.

Lakhina and Eriksen illuminate these requirements with reference to gender, culture, and immigration status, focusing on the climate adaptation and resilience capability of women from refugee backgrounds in the Illawarra region of New South Wales.

Applying this to the topic of climate change adaptation and mitigation, we need to foster new ways of seeing diversity, doing equity work, and being inclusive. This involves:

(i)   Recognizing the individual and collective leadership capabilities of people from marginalized groups;
(ii)  Applying a systems leadership approach that operationalizes intersectionality in a way that seeks to eliminate power imbalances in cross-sector, multi-institutional, trans-border and inter-cultural arenas;
(iii) Enabling people from refugee and migrant backgrounds of the Global South, who are establishing their lives in countries of the Global North, to contribute as thought leaders, structural reformers, and behavioural change agents at the intersection of climate change adaptation and mitigation interventions;

(iv)   Engaging people from refugee and migrant backgrounds in *local* climate change *adaptation* initiatives and in *global* climate change *mitigation* discourse and strategy development—recognizing their unique cross-cultural, cross-country, and cross-community experiences and connections to people and place that embodies lived and learned expertise related to environmental issues, climate conditions, social dimensions, and associated human interactions, spanning the Global South and North; and

(v)   Pursuing climate justice through a human-centred lens where climate change adaptation and mitigation strategies are grounded in decolonizing equity initiatives that transform power structures and relations at local, regional, national, and international levels of political and institutional governance.

This requires courage to explore climate change adaptation and mitigation from different perspectives, value under-represented voices, and interrogate the narratives and cultures we privilege locally, nationally, and globally in climate hazard preparedness, response, and recovery strategies.

## References

Australasian Fire and Emergency Service Authorities Council (AFAC). (2018). *Discussion paper on climate change and the emergency management sector.* https://www.afac.com.au/docs/default-source/doctrine/afac-position-fire-and-emergency-services-and-climate-change.pdf

Australasian Fire and Emergency Service Authority Council (AFAC). (2020). *Valuing differences to enhance SES operational capacity.* https://www.afac.com.au/docs/default-source/publications/afac_valuing-differences_endorsed_2020_11_04.pdf

Berry, K., & Reddy, S. (2010). *Safety with dignity: Integrating community-based protection into humanitarian programming.* Humanitarian Practice Network Paper No. 68.

Lakhina, S. J. (2019). Co-learning disaster resilience: A person-centred approach to understanding experiences of refuge and practices of safety [Doctor of Philosophy thesis. School of Geography and Sustainable Communities, University of Wollongong. https://ro.uow.edu.au/theses1/860

Lakhina, S. J., Eriksen, C., Thompson, J., Aldunate, R., McLaren, J., & Reddy, S. (2019). People from refugee backgrounds contribute to a disaster-resilient Illawarra. *Australian Journal of Emergency Management, 3*(2), 19–20.

New South Wales State Emergency Service. (2020). *Diversity and Inclusion Strategic Framework 2020–2025.* https://www.ses.nsw.gov.au/media/3866/d-i-strategic-framework-2020-25.pdf

UNHCR. (n.d.). *Global compact on refugees.* https://www.unhcr.org/en-au/the-global-compact-on-refugees.html

Young, C., Cormick, C., & Jones, R. (2020). *Learning as we go: Developing effective inclusive management: Case studies and guidance.* Bushfire and Natural Hazards CRC. https://vuir.vu.edu.au/42230/1/Learning%20as%20we%20go%20FINAL-REV-B.pdf

# 3 Moving Away from Climate Crises

## Women's Engagement in Natural Resource Decision Making and Community Monitoring

*Leah Levac, Jane Stinson, and Deborah Stienstra*

**Leah Levac** *is a mother, dog lover, and outdoor enthusiast who lives and works in the traditional territory of the Attawandaron People, and the treaty territory of the Mississaugas of the Credit. She is Associate Professor of Political Science at the University of Guelph, and a faculty advisor of the university's Community Engaged Scholarship Institute. Over the past decade she has been involved in several community-engaged and policy-focused research projects with northern and Indigenous women, focused on their wellbeing in the context of resource extraction.*

*Greater gender equality, economic, and social justice are some of the goals* **Jane Stinson** *has pursued as a community-based researcher with the Canadian Research Institute for the Advancement of Women (CRIAW), and previously as a researcher, educator, and national director with the Canadian Union of Public Employees (CUPE). She is currently Adjunct Research Professor, teaching in the Work and Labour programme of the Institute of Political Economy, Carleton University. She appreciates living on the beautiful but unceded territory of the Algonquin Anishnaabeg Peoples known as Ottawa.*

**Deborah Stienstra** *holds the Jarislowsky Chair in Families and Work at the University of Guelph, where she is Director of the Live Work Well Research Centre and Professor of Political Science. She is the author of* About Canada: Disability Rights *(Fernwood, 2020). Her research and publications explore the intersections of disabilities, gender, childhood, and Indigenousness, identifying barriers to, as well as possibilities for, engagement and transformative change. Her work also contributes to comparative and trans/international research and theory related to intersectional disability rights and justice.*

DOI: 10.4324/9781003089209-6

## Introduction

There are biophysical and socio-cultural links between extractive energy production and climate change. Biophysically, the links between fossil fuel extraction
and consumption, greenhouse gas emissions, and global average temperature
increases are irrefutable (IPCC, 2018). So-called "green" energy projects are
also implicated in this link. For instance, large scale flooding associated with
hydroelectric dams (Calder et al., 2016) is a reminder that even renewable energy
sources contribute to dramatic and dangerous changes to the land and water.
Socio-culturally, one way geographers link climate change and resource extraction is through the concept of environmental displacement, a process through
which communities are prevented from continuing to live on, or access, socio-
cultural resources because the land they have traditionally used has been radically
altered, in some cases driven by desires to extract natural resources (Lunstrum
et al., 2016).

Extractivism is a set of ideas and practices that underpins both the biophysical and socio-cultural links between resource extraction and climate change.
Extractivism refers to a way of thinking and acting that values extracting and selling natural resources for personal and corporate profits with little consideration of
the costs to others and to the environment (Willow, 2016). "Extractivism is...both
a social and an ecological problem" (Willow, 2016, p. 2) where capitalist exploitation of the land, water, and associated resources rationalizes the displacement of
people (Lunstrum et al., 2016). Extractivism is also tightly linked to settler colonialism, an invasive and ongoing structure (Wolfe, 2006) focused on eliminating
the Indigenous *other*, especially to gain territory. The quest to gain territory has
as one of its inevitable consequences the separation of Indigenous Peoples from
their deeply interconnected relationships with, and knowledge of, the land. Settler
colonialism—which Coulthard (2014) argues has become less overtly violent but
nevertheless still aggressively pursued through the politics of recognition[1]—is
also gendered. As Indigenous scholar Joyce Green explains, gendered colonial
processes helped to form stereotypes about Indigenous women, and their objecti-
fication by (primarily male) European settlers created a legacy of marginalization
within the settler colonial society, and in Indigenous ones as well (Green, 2017, p.
5, in Keegan, 2018, pp. 14–15).

The somewhat singular focus of extractivism on securing access to territory
and understanding land as a natural resource to be exploited has another important and pernicious consequence. It displaces and disregards the knowledges and
expertise, not only of Indigenous Peoples (including Indigenous women) who
have been the targets of settler colonialism, but also of others who are often invisible in natural resource decision making, and who may be less likely to benefit
from resource extraction and share capitalist values, and more likely to approach
economic development in different ways. We argue that the knowledges and
expertise of affected populations—often those most invisible in natural resource
decision making—are critical for pushing back against the power of extractivism,
and in turn against its consequences for climate change.

The importance of women's engagement in climate change planning and adaptive responses has been raised by scholars. For instance, in their examination of two international projects related to women's engagement and water management, Figueiredo and Perkins (2013, p. 188) argue that "women have special contributions to make towards climate change adaptation because of gendered differences in positional knowledge and ecological and water-related conditions." Related, though not focused explicitly on women's engagement, others summarize bottom-up approaches to climate change adaptation as featuring several characteristics, including that they: "employ the experience and knowledge of community members to characterize pertinent conditions, community sensitivities, adaptive strategies, and decision making [processes] related to adaptive capacity or resilience" (Smit & Wandel, 2006, p. 285).

Including the experiences and agency of all people means engaging those with disabilities in addressing and responding to climate change, as Bell et al. (2020) argue. This is the first step of applying a disability lens to climate change. In addition, there needs to be a shift from framing women and men with disabilities as victims and as immobile (Bell et al., 2020). Our research (Manning et al., 2018a; Stienstra et al., 2016, 2020) consistently shows the absence of disability analysis in the context of resource extraction. When disability has been part of climate change analyses, attention has often been limited to the disproportionate impacts of natural disasters for people with disabilities due to dependency and lack of mobility, or to a lack of access to services (Lewis & Ballard, 2011); this is important, but offers an incomplete disability analysis. To reframe disability as something other than the creation of vulnerability requires recognizing women and men with disabilities as "active agents of change with valuable knowledge and adaptive capacity" (Albert et al., 2018, p. 2262). This changes the perception of disability and people with disabilities from those who are vulnerable to those who illustrate different ways of living and being in the world. Berghs (2017) suggests people with disabilities help to illustrate biodiversity with the variations of humanity demonstrated through their bodies and diverse ways of being and doing. The implications of the exclusion of people with disabilities in addressing climate change, and responses to address this, have most recently been identified by the Global Action on Disability Network (2021). Some Indigenous Peoples with whom we have worked identify the unique contributions of people with disabilities to their communities, who bring new ways of knowing in relation to community inclusion (Stienstra et al., 2018).

Responses to climate change need to grapple with the pervasiveness of extractivism and be explicitly intersectional. In other words, more sustainability-oriented ways forward depend on better understanding, and learning from, often-invisible community members including Indigenous women, people with disabilities, and others whose knowledges are often overlooked, such as people with low incomes or little formal education, immigrants, migrants and temporary foreign workers, those who identify as LGBTQ2S+, and others who are also often positioned as vulnerable, not as active agents. In this vein, we argue that applying gender-based analysis plus (GBA+)—an intersectionality-informed framework—in impact assessments offers the potential to improve climate change responses through strengthening communities' adaptive capacities, foregrounding Indigenous knowledges, and countering the extractivist mindset which values private profit over addressing widespread economic and social costs. In making this argument, we focus on how northern and Indigenous women's experiences and preferences for engagement in natural resource decision making and community monitoring in present-day Canada can be better reflected in impact assessments, and how more inclusive engagement processes may help to mitigate climate change.

We begin with a brief discussion of intersectionality and critical community engaged scholarship, key theoretical and methodological ideas that have informed our research collaborations over the years. We then present two components of our broader research programmes: A synthesis of our recent work emphasizing the importance of intersectionality and Indigenous women's knowledges in impact assessments, and a case study of a research collaboration that positioned northern and Indigenous women as experts and change agents. Next, we draw on these two areas of work to elaborate what we have learned about integrating intersectionality in impact assessments, first by outlining important principles and then by identifying practices that should be implemented. Finally, we return to our argument that intersectional approaches contribute to mitigating climate change.

## Approaching Our Work: Intersectionality and Critical Community Engaged Scholarship

Our research over the past several years has been underpinned by two important theoretical and methodological orientations: Intersectionality and critical community engaged scholarship. Our understanding of intersectionality, a term coined by Kimberlé Crenshaw (1989), is informed by two recent literature syntheses, which characterize intersectionality as attending to inequality, social context, complexity, relationality, power, and social justice (Collins & Bilge, 2016; Scott & Siltanen, 2017). An intersectional analysis involves understanding how "systems, institutions, and social structures of power intersect with individuals' identities and/or social locations to create temporary or sustained experiences of privilege and exclusion" (Levac & Denis, 2019, p. 4). In our application of intersectionality, we try not to assume a particular criterion as being most important for understanding someone's experience, though most of our work has been with people who identify as women. We also follow those who "acknowledge the possibility of contradictory social positioning (for an individual, a group or a society), and that it is important to consider concurrently *both* the disadvantages (subordination) and advantages (privilege) of relations of inequality, [as well as] the importance of context, in both time and space" (Levac & Denis, 2019, p. 4).

Community engaged scholarship (CES) defies a common definition, but is generally referred to as an orientation to scholarship that minimally [includes] the following characteristics:

(1)  a relationship between the university and its communities;

(2)   a belief that knowledge acquired in the academic setting is strengthened and enhanced by the real world experience found in communities; and

(3)   the power of a mutual, reciprocal, and respectful exchange of ideas, practices, and applications among the engaged partners.

(Whiteford & Strom, 2013, p. 72)

Across disciplines, several shared principles and protocols that inform CES emerge, including reciprocity, responding to community-identified needs, ensuring high-quality investigation (Beaulieu et al., 2018; Mikesell et al., 2013), and self-reflexivity (Wallerstein & Duran, 2008). Drawing on the work of Cynthia Gordon da Cruz (2017) and others, we think about critical community engaged scholarship (CCES) as combining the tenets of CES with commitments to critical race theory and—we suggest—other theories that centre the experiences and knowledges of often-invisible community members, including people with disabilities and Indigenous women. CCES builds on CES by being more explicit about foregrounding power relations in the research process. CCES emphasizes social justice in both the research process and outputs, as well as a commitment to operationalizing research, especially by making sure research results are accessible and available to community members (Gordon da Cruz, 2017).

## Extractivism and the Experiences of Often-Invisible Community Members

Framed by the theoretical and methodological commitments noted above, we now provide a brief synthesis of our recent research focused on the application of intersectionality in natural resource impact assessments, and a case study about a research collaboration in a northern community affected by resource extraction.

### *Synthesis of our Research on (the Lack of) Intersectionality in Impact Assessments*

Since 2015, we have been trying to create a more comprehensive picture of how often-invisible groups within communities are affected by resource extraction, and to identify promising and preferred practices for engaging those who are often overlooked in resource-related planning, decision making, and monitoring. This work has included scoping reviews, policy analyses, and key informant interviews. While our focus has been on the experiences of northern and Indigenous women and girls in present-day Canada, we recognize that the challenges (and opportunities) are global phenomena. In all cases, we have applied an intersectional lens, for instance by asking whose experiences are invisible in literature about resource extraction and its consequences. Throughout the creation of this growing body of research, we have been guided by community advisory groups that have included northern and Indigenous women and/or practitioners with a range of knowledge and experience related to resource development impact assessments. We have also made the results of this work widely available by publishing clear language reports of our findings on websites, hosting public webinars, participating in policy forums, and making formal submissions in policy development processes.

One aspect of this work includes a series of three scoping reviews (Manning et al., 2018a; Stienstra et al., 2016, 2020) conducted between 2016 and 2020—two focused on present-day Canada and one focused internationally.[2] Across these reviews, we examined 360 pieces of academic and community literature. The analytical questions guiding each review varied slightly but all focused on the experiences of people often overlooked and invisible in contexts affected by natural resource extraction. Together, the reviews paint a deeply concerning picture, not only about the experiences of women, but also about a deep lack of understanding of the experiences of other often-invisible communities including people with disabilities, people identifying as LGBTQ2S+, youth, and those at the intersections of these and other exclusions. Literature highlights disproportionately negative consequences for often-invisible communities in areas ranging from employment to education and training, violence, community and social infrastructure, food security, health, housing, and cultural practices. For example, the literature illustrates that impact assessments discount long-term disabling effects of resource extraction and related violence; overlook community infrastructure and service needs; reinforce insecurity of people identifying as LGBTQ2S+; ignore community- and gender-based violence; and disregard links between colonial violence and resource extraction. Bringing these and other issues to light demands hearing from those who are uniquely affected, as we discuss in more detail below. These concerns are prominent in Canadian-focused literature, and evident in international literature, where they are supplemented by concerns about non-Indigenous racialized populations, children and youth, and gendered relations.

A second aspect of this work includes 14 semi-structured key informant interviews with representatives from seven Canadian and international organizations working on natural resource extraction and impact assessment processes (Stienstra et al., 2020), and with seven Indigenous and non-Indigenous women with experience with impact assessments in Canada (Manning et al., 2018b). Both sets of interviews focused on respondents' experiences with—and recommendations for—attending to the experiences of often-invisible communities, and to more inclusive decision making in impact assessments. The insights of these key informants, and of the advisory groups that have helped shape this body of work, have informed our thinking about how impact assessments should be changed to address northern and Indigenous women's experiences and preferences for engagement in natural resource decision making and monitoring.

### *A Case Study of a Community Research Collaboration: Resisting Women's Exclusion with Local Monitoring in Resource-Affected Communities*

This case study illustrates the value of a community engaged research collaboration to support local women to respond to a proponent's disregard of the negative intersectional impacts of resource extraction. Our collaboration centres historically excluded, diverse local women as experts, and puts forward ways of better understanding and monitoring local women's wellbeing in the context of the rapid changes brought about by major resource projects. This case is set in Happy Valley-Goose Bay (HV-GB), Labrador. Labrador, in northeastern Canada, is the traditional territory of both the Inuit and the Innu. A central part of our work together has been the development of a framework to gather locally relevant data on the experiences of women in the community, and particularly those who are often invisible. This collaboration emerged in response to a locally identified need for intersectional baseline data and community monitoring related to the development of a multi-billion-dollar hydroelectric dam nearby at Muskrat Falls on the Lower Churchill River, also known as Mista-Shipu (the great river) in Innu.

With a population of about 8,000, HV-GB is the second-largest town in Labrador. The town is the government and service hub of the region (accessible only by plane or boat from some other parts of Labrador), and home to a diverse population that includes Innu people, Inuit people, and both recent and historic immigrants. The Muskrat Falls hydroelectric project, and a planned project further up-river (Gull Island), are driving population growth and rapid economic change in the town and surrounding area. The geography, cultural diversity, and political context—which consists not only of federal, provincial, and municipal governments, but also Indigenous governments of Nunatsiavut, NunatuKavut, and the Innu Nation (the governments of the Labrador Inuit, NunatuKavut Inuit, and Labrador Innu, respectively)—characterize the complexity of HV-GB.

In 2011, Petrina Beals, then Executive Director of the Mokami Status of Women Council (MSWC), supported by a community-university research alliance (FemNorthNet), made a presentation to the Lower Churchill Joint Review Panel. Beals disputed the claim of the proponent (NALCOR, a provincial crown corporation) that there would be no long-term socio-economic impacts on HV-GB resulting from the proposed hydro development (Hallett & Baikie, 2011, p. 2). She called the Panel's attention to five existing problems—violence against women, poverty, childcare, housing, and mental health, substance abuse and addictions—that were expected to worsen with the influx of highly-paid transient workers, rising costs, and no expansion of social service agencies and government departments to handle the increased need (Hallett & Baikie, 2011, p. 2). The presentation illustrated the willful negligence of an extractivist mentality by NALCOR and the consequences for community-based service agencies:

> Without identifying and acknowledging the truth and having explicit plans and targeted resources, an already stressed social infrastructure will crack – NALCOR and those involved in resource extraction and wealth generation may not look and therefore may not know, but the MSWC and other service agencies will because the casualties of this neglect do, and will, arrive on our doorsteps (Hallett and Baikie, 2011, p. 6).

Beals called for the development of a framework to identify both employment and broader socio-cultural indicators and benchmarks to understand the gendered dimensions of the development; an ongoing, integrated community monitoring process and other mechanisms for meaningful consultation and collaboration to prevent and mitigate negative socio-economic impacts on women in the communities affected by the building of the dam (Hallett & Baikie, 2011, pp. 7–8).

In its final report, the Environmental Assessment Panel acknowledged the needs that Beals had identified. In concluding that "[adverse] effects would be difficult to monitor because of the lack of data and because, by nature, the effects are often hidden" (Canadian Environmental Assessment Agency, 2011, p. xxviii), the Panel suggested that, "mitigation must include a research element" (Canadian Environmental Assessment Agency, 2011, p. xxviii), and recommended that

> the provincial Department of Health and Community Services...conduct a social effects needs assessment, including an appropriately resourced participatory research component, that would determine the parameters to monitor, collect baseline data, and provide recommendations for social effects mitigation measures and an approach to on-going monitoring.
>
> (Canadian Environmental Assessment Agency, 2011, p. 291)

This critical—unheeded—call by the Environmental Assessment Panel served as the foundation for developing a new collaborative research initiative to develop what came to be known as the *Community Vitality Index* (CVI), a framework with corresponding survey and community data analysis focused on informing monitoring efforts with local data about women's experiences.

The development of the CVI was guided by intersectionality and CCES. The process to develop the CVI was both iterative and flexible, evolving with the understanding of women's wellbeing in HV-GB. A group of diverse local women were engaged in defining wellbeing, and in creating a framework, survey, and indicators to monitor wellbeing. As a result of their active engagement, the process developed local women's capacity for leadership and advocacy. Local women served as advisors and supported the implementation of all aspects of the project. In total, 26 Indigenous and non-Indigenous women between the ages of 18 and 60 with diverse family compositions, physical and mental abilities, sexual orientations, and educational backgrounds participated in one of two, two-day workshops that created the first version of the CVI (Levac et al., under review). The process of creating the CVI did not set out to link specific participants' contributions to their identities, though this did happen in some instances. Instead, the focus was how the group's diversity and rich interactions generated outputs (the framework, survey, and indicators) that account for the complexity of diverse women's wellbeing. Several inclusion strategies were used in the design and implementation of the workshops to ensure the inclusion of women who were often not included in community consultations. These included holding events in physically accessible locations; providing a stipend, meals, transportation, and childcare; and providing a separate room where women could take a break, regain energy, or speak with a support person if needed. These measures were informed by the literature review and our collective experience of how to put intersectionality into practice.

Five components of wellbeing—physical, emotional, spiritual, mental/intellectual, and cultural—were identified, defined, and illustrated with images reflecting the local culture and women's conceptions of wellbeing. We developed a comprehensive survey to collect quantitative and qualitative data about the community's wellbeing, using questions based on participants' contributions as well as wellbeing surveys developed in other jurisdictions. Community participants were trained to conduct the survey with others and facilitated subsequent stages of data collection where survey data were collected both in-person and online. This step not only helped to refine the survey, but it also fostered local women's leadership through their deep participation (Ospina & Foldy, 2010, p. 300).

The participatory method of constructing definitions and measures of wellbeing led to several differences from standard definitions of wellbeing. For example, participants highlighted how their wellbeing was dependent on environmental sustainability and how their cultural wellbeing was deeply connected to the environment (Levac et al., under review). Although, as noted above, women's contributions were synthesized to produce rich and nuanced outputs, it was clear that women's unique identities lent expertise to the process. For instance, women participants living with low-income pointed out ways that available services were not accessible to them, and how this compromised their wellbeing. Women with mental health challenges highlighted how difficult it was to obtain mental health services locally, again to the detriment of their wellbeing. Women facing chronic stress because of their exclusion—often related to features of their identities—articulated the importance of understanding behaviours such as gambling, smoking, and drinking as coping mechanisms, rather than as "poor health behaviours."

The women who were involved in developing the CVI participated not only in a collaborative research process, but also in a relational leadership development process that allowed them to grow by connecting with other women, hearing each other's stories, and finding their voices by being heard (Levac et al., under review). This case highlights the valuable contributions women with diverse identities can make when they are positioned and recognized as experts, in this case to identify how building a major dam will affect the health, wellbeing, and environmental sustainability of their community. The data generated through the CVI was collaboratively analyzed by the research team and community members. Community leaders and policy makers interested in following up on the recommendations of the Environmental Assessment panel to monitor and address the impacts associated with the development of the dam will find the data useful. Community organizations and Indigenous and settler governments and agencies have also expressed interest in the data as a way of informing their work. Also, the CVI process provides a model for using intersectionality and CCES in the development of community wellbeing indicators that could be applied elsewhere.

## Pushing Against the Ideas and Practices of Extractivism

Our research—both in HV-GB and focused on synthesizing and creating knowledge about intersectionality in impact assessments—has revealed persistent, primarily negative, impacts of resource extraction for often-invisible community members such as Indigenous women, LGBTQ2S+ persons, people with disabilities, and youth. As we have described above, these impacts range from violence, to disrupted food sources, to the exaggeration of chronic stressors and overextension of community services. People with whom we have collaborated, and groups identified in the literature we have reviewed, have also historically been excluded from public engagement processes, and invisible in impact assessments, creating an engagement gap with serious consequences. For instance, literature suggests that when the impacts of pollution on pregnant women are ignored, or the violence resulting from resource extraction activities is unaddressed, the long-term disabling effects of resource extraction and related violence are overlooked, and future community service and infrastructure needs are invisible. When binary understandings of gender persist, queer and transgender people remain invisible. In turn, community- and gender-based violence in resource-affected communities, combined with significant threats to personal safety faced by those identifying as LGBTQ2S+, make it likely that LGBTQ2S+ persons are at extreme risk for violence in resource extractive contexts.

Fortunately, policy development processes related to resource extraction, including impact assessments, have the potential of offering many opportunities for broader engagement by the public, communities, groups, and individuals. Unfortunately, these processes are often exclusive. They require technical knowledge, are abstracted from daily life, are often limited to elite community members, unfold according to short timelines, and offer limited opportunities for participant capacity building. In other words, the processes themselves are also driven by extractivist logics that privilege efficiency and undermine inclusion. However, from our research described in this chapter, we have learned about principles and practices of engagement that are more inclusive of those groups and individuals who are often invisible and marginal in resource extraction-related policy development.

### *Guiding Principles for Inclusive Engagement*

Elsewhere, based on our literature syntheses and key informant interviews, we have detailed guiding principles that should underpin every stage of decision making related to resource extraction, particularly in the process of conducting impact assessments (Manning et al., 2018b). These include commitments to recognizing, valuing, and incorporating Indigenous knowledges and the expertise of Indigenous women; ensuring access and inclusion to processes, including by requiring GBA+ as part of understanding and responding to project impacts; providing sufficient resources to facilitate not only basic engagement but also broader capacity-building initiatives; and recognizing that resistance—especially through the work of land defenders—is a legitimate form of engagement (Manning et al., 2018b, pp. 39–45). Here, we adapt and expand on these principles to consider more carefully what we have learned about taking up intersectionality in impact assessments, and about the possibilities for resisting extractivism.

*Recognize, value, and incorporate Indigenous knowledges and expertise.* This requires seeing Indigenous knowledge holders as experts, supporting Indigenous-led research using Indigenous methodologies, and prioritizing Indigenous co-management of impact assessment processes (Manning et al., 2018b, pp. 39–40). While these commitments are essential for ensuring that resource extraction processes begin to redress some of the effects of settler colonialism on Indigenous Peoples, they do not change the reality that policy making around resource extraction remains guided by settler, rather than Indigenous, customs and laws. Further, recognition (of knowledges and expertise), as Coulthard (2014) notes, does not necessarily lead to the return of stolen land. To more fully counter extractivist logics, a commitment to Indigenous knowledges and expertise demands a commensurate commitment to broader sovereignty-related reforms.

*Recognize, value, and incorporate the expertise of under-represented groups of women.* Our research collaborations demonstrate significant gendered impacts of resource extraction and related gendered gaps in policy making. Not only has settler colonialism devalued and objectified Indigenous women, resulting in their marginalization within settler societies, but many women and non-binary persons remain invisible, despite having knowledges resulting—as Figueiredo and Perkins (2013) argue—from their locations and roles in communities. Both the disproportionately negative impacts and the engagement gaps need to be addressed. This is accomplished by ensuring under-represented women, including Indigenous women, are represented and engaged in significant numbers at every stage of the process.

*Commit to intersectional analyses by taking up GBA+.* Understanding GBA+ as a framework for undertaking intersectional analyses helps to identify and address—both conceptually and practically—systemic inequities and assumptions about the benefits and burdens that result from public policies and programmes. When used throughout the impact assessment process (e.g., while developing baseline data, making project-related decisions, and undertaking monitoring, evaluation, and mitigation) an intersectional analysis can identify and mitigate concerns of those often invisible in communities, including Indigenous women. Any GBA+ or intersectional analysis needs to be culturally relevant; an example is the Inuit-specific gender-based analysis framework developed by Pauktuutit, the national Inuit women's organization (Pauktuutit et al., 2012). Importantly, and as noted above in our discussion about our theoretical and methodological research approach, an intersectional analysis goes beyond gender to recognize differential impacts on, and needs of, different groups including women and men with disabilities, those in the LGBTQ2S+ communities, among many others.

*Ensure access and inclusion.* Access is more than a word or a checklist to be completed. Access is a process, perception, and a set of relationships essential to both inclusion and participation. Titchkosky argues that

access, then, is tied to the social organization of participation, even to belonging. Access not only needs to be sought out and fought for, legally secured,
physically measured, and politically protected, it also needs to be understood—as a complex form of perception that organizes socio-political relations between people in social space.

(Titchkosky, 2011, p. 4)

Access means creating processes that structure often-invisible community members into the discussion from the beginning. These processes of inclusion then
help to shift the perceptions of whose knowledge is valued and who is relevant.

*Recognize resistance.* Resistance, land defense actions, protest, and legal
actions taken in response to resource extraction and development are important
signals by Indigenous people and others. These actions communicate frustrations
with inadequate consultation, and concerns about dangerous outcomes, inadequate
monitoring, and insufficient enforcement of mitigation measures. Resistance indicates problems with engagement and decision-making processes that governments and proponents need to address (Manning et al., 2018b, p. 44). It is an
important reminder to the Canadian state and project proponents that Canada is a
signatory to the United Nations Declaration on the Rights of Indigenous Peoples
(UNDRIP), which includes the principle of free, prior, and informed consent
(FPIC), defined as

[the] consensus of all members of the [Indigenous community]...determined
in accordance with their customary laws and practices free from any external
manipulation, interference and coercion and obtained after fully disclosing
the intent and scope of the [project] in a language and process understandable
to the community.

(in Minter et al., 2012, p. 1242)

Applying the principles of FPIC to the impact assessment process could lead to
different outcomes and less resistance.

### Practices that Support Inclusive Engagement

Adopting principles that support inclusive engagement is an important foundational step for advancing intersectional analysis in impact assessments to counter
the logics of extractivism and, in turn, the impacts of climate change. However,
these principles must be supported by specific practices that can help to not only
structure often-invisible communities and groups into important policy conversations, but also ensure that the right supports are in place to make participation
possible.

*Structuring often-invisible communities into the process.* A rights-based
approach to impact assessment, grounded in international human rights commitments, is one way to set the stage for ensuring often-invisible communities are
included. Integrating commitments from UNDRIP as part of resource extraction

policy frameworks provides an opportunity to recognize and acknowledge the legacy of colonialism on Indigenous women, to respect the rights of Indigenous Peoples, and to involve Indigenous women in all levels of decision making (Manning et al., 2018b, p. 45). A similar approach can be used to recognize and value the contributions of others by including commitments made through other UN conventions, such as those on the Rights of Persons with Disabilities, and on the Rights of the Child.

Another way to facilitate the meaningful engagement of often-invisible communities is to identify and remove the structural barriers that currently make engagement difficult. Generally speaking, opportunities for engaging in resource extraction decision making often rely on substantial technical expertise or knowledge as well as timely awareness of and participation in these opportunities. Indigenous women's voices are often missing from engagement processes. Information about impact assessment processes is not always accessible in multiple formats, or in ways that enable community members with disabilities to engage effectively. Key informants across our research have articulated important procedural changes that are necessary, including: Changing the formal and adversarial style of impact assessment hearings to make them less intimidating, ensuring adequate funding for participation, ensuring access (e.g., through the provision of childcare, transportation, sign language interpreters, etc.) (Manning et al., 2018b, p. 64), "allowing sufficient time to respond to government and proponent reports and requests, and facilitating capacity building in the area of impact assessment for Indigenous women and their organizations" (Manning et al., 2018b, p. 64). Capacity building could include community workshops about the impact assessment process, or support with public speaking skills. Key informants also pointed out the need for Indigenous women to be more involved in community monitoring of project impacts; developing worksite training, policies, and procedures related to cultural protocols, inclusion, and safety; and developing mitigation strategies to address the consequences of resource extraction (Manning et al., 2018b). Here, the tenets of CCES, including responding to community-identified needs and foregrounding power relations, offer important guidance.

A third practice aimed at improving the inclusion of often-invisible communities is to ensure that appropriate data are available to inform the impact assessment. In other words, it is very difficult to determine which impacts will disproportionately affect whom if community data are not available. Data on impacts should be disaggregated by gender and other aspects of identity, and should be locally relevant. The case study discussed above illuminates why this is important and how the creation of locally relevant data can proceed. The creation of locally relevant and disaggregated data can be facilitated by engaging with often-invisible communities and the organizations that serve them, including Indigenous women, women with disabilities, young women, and others. Useful data can guide the development of impact statements and mitigation plans, and proponents' and governments' understanding of long-term cumulative impacts that may affect different communities. Disaggregated data are useful for uncovering intersections of impact, such as the long-term impacts on child-bearing women of eating seal contaminated by methylmercury (Manning et al., 2018b, p. 54) or the impacts of gender-based violence on women with disabilities and those that identify as LGBTQ2S+ and their ability to access services in the future (Levac et al., 2021).

*Ensuring the availability of adequate and appropriate supports.* To complement the creation of structures that invite often-invisible communities into the process, another important practice is to ensure that adequate and appropriate supports are available for enabling participation. Many of these have been discussed above.

Across our research, we have also learned that positive actions can be taken to engage under-represented groups of women. These include holding women-only consultation sessions; providing interpretation services and clear-language explanations of processes, technical reports, and materials; working with community members to design materials and tools appropriate for that community; requiring seats for Indigenous women on impact assessment panels; holding meetings in accessible locations; and providing support people who may include sign language interpreters, attendant care providers, among others. Ultimately, if women are to participate, they will need adequate funding to prepare submissions and to facilitate their participation (Manning et al., 2018b, pp. 44, 51).

*Supporting community-led consultations and community-based impact assessments.* Finally, ensuring under-represented women's full participation and leadership can be facilitated through community-led consultations, community-based impact assessments (CBIAs), and other community-led forms of impact assessment and monitoring. As we argue elsewhere (Levac et al., 2021), community-led consultations and CBIAs offer promise because they align with the principles of intersectionality and CCES and enable practices that can redress power inequities. Examples of community-led consultations and CBIAs can be found around the world, including in Guatemala, Sweden, Australia, and Canada (Levac et al., 2021). Shifting the focus of, and responsibility for, resource extraction engagement practices from the government and proponents to affected communities is an essential part of challenging extractivism. As we note in the above case study, and as is evident in international examples (Levac et al., 2021), community-based consultations, CBIAs, and community monitoring contribute critical perspectives and knowledge to the process and can engage more community members than is often possible through proponent-led consultations.

CBIAs, which cannot be done without sufficient funding and support, can include community members through workshops, meetings, interviews, and participatory mapping, creating a community baseline, identifying potential effects, and monitoring related assessment processes (Levac et al., 2021). The social water assessment protocol (SWAP), developed in Australia, is one example of a CBIA. Grounded in a human rights approach to water, the SWAP is a scoping tool to identify water-related impacts that are experienced by diverse members of communities. Intended to complement existing water-related assessment requirements in traditional environmental impact assessments, its 14 themes capture a wide range of impacts for different members of communities. Themes—including uses of water; interactions and significance of water to Indigenous Peoples; spiritual and cultural values of water; and health issues related to water—are accompanied by guiding questions to be answered using many different sources of information. A SWAP can reveal areas where a more in-depth social or human rights impact assessment is required (Collins & Woodley, 2013). The SWAP calls for attention to "the current and historical interactions between Indigenous Peoples and government, mining companies, and other stakeholders" (Collins & Woodley, 2013, p. 163), which recognizes how persistent marginalization over time—through colonialism and ableism, for instance—remains relevant in addressing community needs moving forward.

Community-based monitoring (CBM) is another way to meaningfully engage with the knowledges of often-invisible communities. CBM is "a process of routinely observing environmental or social phenomena, or both, that is led and undertaken by community members and can involve external collaboration and support of visiting researchers and government agencies" (Johnson et al., 2015, p. 2). CBM, which depends on extensive community involvement, offers several benefits. For example, it improves access to monitoring in remote and isolated locations and draws on deep, intergenerational local and Indigenous knowledges (Johnson et al., 2015).

## Conclusion: Intersectionality, Inclusive Engagement, and Resisting Extractivism

There are clear ways forward for more carefully attending to the experiences and knowledges of often-invisible communities and community members in resource extraction-related policy decisions. We argue that the meaningful inclusion of under-represented women, facilitated by applying the intersectionality-informed framework of GBA+, can improve our collective response to climate change. The principles we highlight above provide a strong foundation for this inclusion, by offering guidance for taking up intersectionality and CCES in impact assessments. The principles can be thought of as values or commitments that, when taken seriously, re-shape how we think about the politics of resource extraction, including whose knowledges matter, and which people are invited to shape the decisions being made. The principles are accompanied by a set of practices. The practices provide actionable guidance on what it actually looks like to facilitate the inclusion of often-invisible communities. Like the principles, the practices highlight not only what we have learned through the body of research and case study discussed earlier, but also our underlying commitments to intersectionality and CCES.

Taking up these principles and practices can help to address climate change by challenging extractivism in several ways. First, their adoption by proponents would disrupt the extractivist mentality by slowing down the decision-making process and allowing for broader and more meaningful consultations, hopefully leading to decisions based on non-extractivist values. For instance, an impact assessment process grounded by Indigenous knowledges and values (which often include recognizing land as a life and knowledge source) challenges and disrupts the settler colonial and extractivist view of land as a natural resource to exploit for private profit.

Second, including intersectional considerations that favour equity and recognize the importance of long-term environmental sustainability can challenge the often-foregrounded benefit of temporary job creation. An intersectional analysis helps to reveal how benefits, as well as the consequences of lasting environmental damage, are likely to be unevenly distributed, such that people least likely to benefit are also most likely to shoulder the associated burdens. Correcting these inequities could help to mitigate activities that advance climate change.

Third, more inclusive planning processes and sensitivity to gender and other differences can help communities strengthen their adaptive capacity for dealing with environmental and climate changes. If impact assessment processes were guided by the above principles and practices, they could help communities plan for and deal with environmental and climate changes and specifically address women's greater vulnerability to climate change risks by increasing their inclusion in planning and decision-making processes (Reed et al., 2014, p. 996).

While our research findings make us strong advocates for meaningful engagement, we recognize that wider spread public participation does not come without challenges. One tension arises between engagement and the principle of recognizing resistance. Repeated government and proponent failures to make substantive shifts towards new decision-making models have understandably led some people to decide that engagement inevitably leads to co-optation and the loss of rights, a concern that we take seriously. However, we contend that both/and thinking is useful here. In other words, and in keeping with our CCES commitments, recognizing resistance while concurrently building bridges between communities and institutions is important for moving forward on more equitable futures. Related, it is also not appropriate to assume that all engagement leads to resisting resource extraction. We acknowledge that many Indigenous Nations support—in various ways—resource development projects in their territories. Our argument for more fulsome inclusion—informed by intersectionality—is not meant to suggest that everyone will agree, but that groups overlooked in a community because of power imbalances and inequities should also have meaningful opportunities to identify negative consequences and potential benefits. The purpose is to develop plans that take the needs of frequently overlooked members of the community into account, to ensure a more equitable distribution of benefits and burdens.

Another engagement-related challenge involves the geopolitical and social contexts in which engagements take place. For instance, these contexts have had a dramatic impact on how the community research collaboration described above has unfolded. Competing land claims, disagreements between Indigenous groups as to the benefits and burdens of resource projects, governance instability within local organizations, the broader context of settler colonialism, individual and sub-national economic disparities, and other factors have radically shaped women's preferences for engagement. These complex dynamics point to the importance of flexibility, and of being able to adjust the practices of engagement to meet changing community needs and preferences. Finally, as we have alluded to throughout our discussion of principles and practices, communities need resources to help facilitate both capacity-building and widespread inclusive engagement, and such resource provision has historically been inadequate in the Canadian context.

On this point, and more generally related to the uptake of these principles and practices, there is reason for optimism. In 2019, Canada adopted a new Impact Assessment Act (IAA, 2019), which—among other things—includes important provisions for undertaking more inclusive impact assessments. These provisions offer the potential for broadening public engagement in decisions about resource extraction projects, in turn influencing the final outcomes. For example, more attention is paid to considering impacts on Indigenous Peoples, and to using Indigenous knowledges to inform impact assessments (IAA, 2019, secs. 22(1) c, g & 119). The Act also frames consultation with Indigenous Peoples based on Indigenous rights and the principle of FPIC (IAA, 2019), although discrepancies between the Government of Canada's understanding of, and UNDRIP's definition of, FPIC may prove problematic. Further, the Act introduces a new planning phase intended to determine whether a project should go forward, whether to carry out an assessment, and if so, what the scope of the assessment should be (Fasken, 2019). Further, the shift of the Act's title from the Canadian Environmental Assessment Act to the Impact Assessment Act signals its intention to be more inclusive of social impacts, including those related to social determinants of health, as well as environmental and economic impacts over the long term. Finally, the Act insists on the consideration of "the intersection of sex and gender with other identity factors" (IAA, 2019, sec. 22 (1) s) as one of the mandatory factors for consideration in determining impacts.

Communities may benefit from these legislative requirements, from modest funds available from the Impact Assessment Agency of Canada (IAAC) to support inclusive engagement, and from ongoing work to develop guidelines related to implementing GBA+ in impact assessments (IAAC, 2021). Still, the timelines for impact assessments remain tight. For example, the IAAC must decide whether to conduct an assessment within 180 days of posting a project description online. The principles and practices described above imply that longer timelines are required, especially when it is necessary to concurrently address deeply rooted structures of exclusion. It takes time, a wide range of approaches and strategies, and many iterative discussions, for often-invisible community members to learn of and deeply consider a project, apply intersectional principles and practices, and identify the likely impacts it will have on them. Countering extractivist logics—so central to "capitalist penetration and related displacement" (Lunstrum et al., 2016, p. 132)—promises to be a long but worthwhile journey. Taking seriously diverse northern and Indigenous women's experiences and preferences for policy engagement via the creation of more inclusive engagement principles and practices can challenge extractivism, and, in turn, help to mitigate climate change.

## Notes

1  In *Red Skin, White Masks: Rejecting the Colonial Politics of Recognition*, Coulthard (2014) argues that the recognition of Indigenous cultural rights in present-day Canada still takes place within the confines of the settler state, therefore understanding Indigenous rights only within the confines of the colonial, racist, and capitalist features of liberal democracies. "Recognition," offered in this context, helps to obscure the focus on the critical issue of returning land.
2  Included countries were Australia, Bolivia, Brazil, Colombia, the Democratic Republic of the Congo, Ecuador, Ghana, Guatemala, Honduras, India, Mexico, Norway, Panama, Papua New Guinea, Philippines, Peru, South Africa, Sweden, and the United States.

## References

Albert, S., Bronen, R., Tooler, N., Leon, J., Yee, D., Ash, J., Boseto, D., & Grinham, A. (2018). Heading for the hills: Climate-driven community relocations in the Solomon Islands and Alaska provide insight for a 1.5 C future. *Regional Environmental Change*, *18*, 2261–2271.

Beaulieu, A., Breton, M. & Brousselle, A. (2018). Conceptualizing 20 years of engaged scholarship: A scoping review. *PLoS ONE, 13*, e0193201.

Bell, S. L., Tabe, T., & Bell, S. (2020). Seeking a disability lens within climate change migration discourses, policies and practices. *Disability & Society, 35*(4), 682–687.

Berghs, M. (2017). Practices and discourses of Ubuntu: Implications for an African model of disability? *African Journal of Disability, 6*(1), 8. https://doi.org/10.4102/ajod.v6i0.292

Calder, R. S. D., Schartup, A. T., Li, M., Vlaberg, A. P., Balmcom, P. H., & Suderland, E. M. (2016). Future impacts of hydroelectric power development on methylmercury exposures of Canadian Indigenous communities. *Enviromental Scence & Technology, 50*(23), 13115–13122. https://pubs.acs.org/doi/abs/10.1021/acs.est.6b04447

Canadian Environmental Assessment Agency, Lower Churchill Hydroelectric Generation Project Joint Review Panel, Newfoundland and Labrador Department of Environment and Conservation, Government of Newfoundland and Labrador, & Intergovernmental Affairs Secretariat. (2011). *Report of the joint review panel: Lower Churchill hydroelectric generation project, Nalcor energy, newfoundland and labrador*. Ottawa: Canadian Environmental Assessment Agency.

Collins, P. & Bilge, S. (2016). *Intersectionality*. Polity Press.

Collins, N., & Woodley, A. (2013). Social water assessment protocol: A step towards connecting mining, water and human rights. *Impact Assessment & Project Appraisal, 31*(2), 158–167. doi:10.1080/14615517.2013.774717.

Coulthard, G. (2014). *Red skin, white mask: Rejecting the colonial politics of recognition*. University of Minnesota Press.

Crenshaw, K. (1989). Demarginalizing the intersection of race and sex: A Black feminist critique of antidiscrimination doctrine, feminist theory and antiracist politics. *The University of Chicago Legal Forum, 1989*(1), 139–167.

Fasken. (2019). The New Federal Impact Assessment Act. *Environmental Bulletin*. https://www.fasken.com/en/knowledge/2019/08/the-new-federal-impact-assessment-act/

Figueiredo, P., & Perkins, P. E. (2013). Women and water management in times of climate change: Participatory and inclusive processes. *Journal of Cleaner Production, 60*(2013), 188–194.

Global Action on Disability Network (GLAD). (2021). *Promoting disability-inclusive climate change action*. Global Action on Disability Network Secretariat. https://gladnetwork.net/search/resources/secretariat-global-action-disability-glad-network-presents-issue-paper-and-guide

Gordon da Cruz, S. (2017). Critical community-engaged scholarship: Communities and universities striving for racial justice. *Peabody Journal of Education, 93*(3), 363–384.

Green, J. (2017). Taking more account of Indigenous feminism [Introduction]. In J. Green (Ed.), *Making sense of indigenous feminism* (2nd ed., pp. 1–20). Fernwood.

Hallett, V., & Baikie, G. (2011). *Out of the rhetoric and into the reality of local women's lives*. Submission to the Environmental Assessment Panel on the Lower Churchill Hydro Development by the Mokami Status of Women Council with FemNorthNet. https://www.criaw-icref.ca/publications/out-of-the-rhetoric-and-into-the-reality-of-local-womens-lives/

Impact Assessment Act (IAA) 2019 (S.C.) c. 28, s. 1. (2019). https://laws-lois.justice.gc.ca/PDF/I-2.75.pdf

Impact Assessment Agency of Canada (IAAC). (2021). *Guidance: Gender based analysis plus in impact assessment*. Impact Assessment Agency of Canada, Government of Canada. https://www.canada.ca/en/impact-assessment-agency/services/policy-guidance/practitioners-guide-impact-assessment-act/gender-based-analysis.html

IPCC. (2018). Summary for policymakers. In V. Masson-Delmotte, P. Zhai, H. -O. Pörtner, D. Roberts, J. Skea, P. R. Shukla, A. Pirani, W. Moufouma-Okia, C. Péan, R. Pidcock, S. Connors, J. B. R. Matthews, Y. Chen, X. Zhou, M. I. Gomis, E. Lonnoy, T. Maycock, M. Tignor, & T. Waterfield (Eds.), *Global warming of 1.5°C. An IPCC special report on the impacts of global warming of 1.5°C above pre-industrial levels and related global greenhouse gas emission pathways, in the context of strengthening the global response to the threat of climate change, sustainable development, and efforts to eradicate poverty*. https://www.ipcc.ch/site/assets/uploads/sites/2/2019/05/SR15_SPM_version_report_LR.pdf

Johnson, N., Alessa, L., Behe, C, Danielsen, F., Gearheard, S., Gofman-Wallingford, V., Kliskey, A., Krümmel, E., Lynch, A., Mustonen, P., & Svoboda, M. (2015). The contributions of community-based monitoring and traditional knowledge to arctic observing networks: Reflections on the state of the field. *Arctic, 68*(Suppl 1), 1–13.

Keegan, A. (2018). *Framing of violence against Indigenous women in Canadian House of Commons discourse*. [Unpublished Master's major research paper, University of Guelph].

Levac, L., & Denis, A. (2019). Combining feminist intersectional and community-engaged research commitments: Adaptations for scoping reviews and secondary analyses of national data sets. *Gateways: International Journal of Community Research and Engagement, 12*(1), 1–19. https://doi.org/10.5130/ijcre.v12i1.6193

Levac, L., Stienstra, D., Beals, P., & McCuaig, J. (under review). Advancing intersectional considerations in measuring gender equality: The participatory development of a Community Vitality Index in Happy Valley-Goose Bay, Labrador. In C. Gabriel & P. Rankin (Eds.), *Critical Perspectives on Gender Equality Measurement in Canada*. UBC Press.

Levac, L., Stinson, J., Manning, S.M, & Stienstra, D. (2021). Expanding evidence and expertise in impact assessment: Informing Canadian public policy with the knowledges of invisible communities. *Impact Assessment and Project Appraisal, 39*(3), 218–228. https://doi.org/10.1010/14615517.2021.1906152

Lewis, D., & Ballard, K. (2011). Disability and climate change: Understanding vulnerability and building resilience in a changing world. In Pamela Thomas (ed.), *Development Bulletin: Implementing Disability-Inclusive Development in the Pacific and Asia, 74.*

Lunstrum, E., Bose, P., & Zalik, A. (2016). Environmental displacement: The common ground of climate change, extraction and conservation. *Area, 48*(2), 130–133. https://doi.org/10.1111/area.12193

Manning, S., Nash, P., Levac, L., Stienstra, D., & Stinson, J. (2018a). *A literature synthesis report on the impacts of resource extraction for Indigenous women.* Canadian Research Institute for the Advancement of Women (CRIAW). https://www.criaw-icref.ca/wp-content/uploads/2021/04/Impacts-of-Resource-Extraction-for-Indigenous-Women.pdf

Manning, S., Nash, P., Levac, L., Stienstra, D., & Stinson, J. (2018b). *Strengthening impact assessments for Indigenous women.* Canadian Research Institute for the Advancement of Women (CRIAW). https://www.criaw-icref.ca/images/userfiles/files/FINAL_CEAAReport_Dec7.pdf

Mikesell, L., Bromley, E., & Khodyakov, D. (2013). Ethical community-engaged research: A literature review. *American Journal of Public Health, 103*(12), e7–e14.

Minter, T., de Brabander, V., van der Ploeg, J., Persoon, G. A., & Sunderland, T. (2012). Whose consent? Hunter-gatherers and extractive industries in the northeastern Philippines. *Society & Natural Resources, 2*(12), 1241–1257. https://doi.org/10.1080/08941920.2012.676160

Ospina, S., & Foldy, E. (2010). Building bridges from the margins: The work of leadership in social change organizations. *The Leadership Quarterly, 21*(2), 292–307. https://doi.org/10.1016/j.leaqua.2010.01.008

Pauktuutit Inuit Women of Canada, Rasmussen, D., & Guillou, J. (2012). Developing an Inuit-specific framework for culturally relevant health indicators incorporating gender-based analysis. *Journal of Aboriginal Health, 8*(2), 24–35.

Reed, M. G., Scott, A., Natcher, D., & Johnston, M. (2014). Linking gender, climate change, adaptive capacity, and forest-based communities in Canada. *Canadian Journal of Forest Research, 44,* 995–1004. dx.doi.org/10.1139/cjfr-2014-0174

Scott, N. & Siltanen, J. (2017). Intersectionality and quantitative methods: Assessing regression from a feminist perspective. *International Journal of Social Research Methodology, 20*(4), 373–385. https://doi.org/10.1080/13645579.2016.1201328

Smit, B., & Wandel, J. (2006). Adaptation, adaptive capacity and vulnerability. *Global Environmental Change, 16*(3), 282–292. https://doi.org/10.1016/j.gloenvcha.2006.03.008

Stienstra, D., Levac, L., Baikie, G., Stinson, J., Clow, B., & Manning, S. (2016). *Gendered and intersectional implications of energy and resource extraction in resource-based communities in Canada's North.* Canadian Research Institute for the Advancement of Women.

Stienstra, D., Baikie, G., & Manning, S. (2018). "My granddaughter doesn't know she has disabilities and we are not going to tell her": Navigating intersections of Indigenousness, disability and gender in Labrador. *Disability and the Global South, 5*(2), 1385–1406.

Stienstra, D., Manning, S.M, & Levac, L. (2020). *More promise than practice: GBA+, intersectionality and impact assessments.* https://liveworkwell.ca/sites/default/files/pageuploads/Report_Mar31_AODA.pdf

Titchkosky, T. (2011). *The question of access: Disability, space, meaning.* University of Toronto Press.

Wallerstein, N., & Duran, B. (2008). The theoretical, historical, and practice roots of CBPR. In M. Minkler & N. Wallerstein (Eds.), *Community based participatory research for health: From process to outcomes* (2nd ed., pp. 25–45). Jossey Bass.

Whiteford, L., & Strom, E. (2013). Building community engagement and public scholarship into the university. *Annals of Anthropological Practice, 37*(1), 72–89.

Willow, A. J. (2016). Indigenous extrACTIVISM in Boreal Canada: Colonial legacies, contemporary struggles and sovereign futures. *Humanities, 5*(55), 1–15. https://doi.org /10.3390/h5030055

Wolfe, P. (2006). Settler colonialism and the elimination of the native. *Journal of Genocide Research, 8*(4), 387–409. https://doi.org/10.1080/14623520601056240

# Reflection on Chapter 3

## Anchoring the Hope: Decision-making Safeguards to Make Women's Voices Count

*Anna Johnston*

**Anna Johnston** *is a staff lawyer at West Coast Environmental Law, where she works on environmental impact assessment, cumulative effects, constitutional and climate law. She co-chairs the Environmental Planning and Assessment Caucus of the Réseau-Canadian Environmental Network, and in 2020 was appointed to the federal Minister's Advisory Council on impact assessment. From 2016–2019 she was a delegate to the Multi-Interest Advisory Committee appointed to assist in the review of Canada's federal environmental impact assessment processes. Anna holds a Master of Laws from Dalhousie University, and an LLB and BA from the University of Victoria.*

Levac, Stinson, and Stienstra's chapter is both encouraging and dismaying. It is encouraging because it illustrates a growing recognition of the unequal distribution of climate change-related impacts, benefits, risks, and the uncertainties of projects along various identity lines, and awareness of the need to apply an intersectional lens to resource planning and decision making. But it is discouraging because, as the chapter suggests, environmental decision making in Canada (and throughout the world) tends to exclude the voices, knowledge, and authority of those who often stand to be most affected. So long as decision making remains highly discretionary, is based on a subjective "balancing" of competing interests that include short-term economic interests, rarely if ever meaningfully considers "no" as an option, and is made unilaterally by state authorities (often perpetuating settler colonialism), it is destined to fail to achieve the authors' vision. For that reason, I do not share their optimism that the *Impact Assessment Act* (IAA) as currently written will lead to more equitable, inclusive, and informed decisions that help Canada live up to its international climate commitments or protect its most vulnerable citizens from the effects of climate change.

Environmental impact assessment (EIA) has a long history in Canada. First required of federal departments in 1973 (FEARO, 1987), EIA has been informing federal decisions for decades. It emerged in response to a growing awareness in

DOI: 10.4324/9781003089209-7

the 1960s of the inadequacy of the then-fragmented approach to environmental decision making, including its inability to consider the full range of perspectives and information on proposed undertakings necessary to make informed, fair decisions that advance sustainability (Winfield, 2016). However, nearly fifty years of experience in Canada has shown that EIA is not living up to its full potential (Gibson et al., 2016).

In particular, EIA has overwhelmingly been used as a process for making projects a little less harmful and for finding justification for residual adverse impacts, regardless of their distribution. Under both the original *Canadian Environmental Assessment Act* (CEAA) and the *Canadian Environmental Assessment Act, 2012* (CEAA, 2012), responsible authorities had to determine whether projects would result in significant adverse environmental effects and if so, whether those effects were "justified in the circumstances" (CEAA, ss 23(1), 37(1); CEAA, 2012, s 52). Only rarely (such as with the Prosperity Gold-Copper Mine Project in British Columbia, as well as its second iteration, the New Prosperity Gold-Copper Mine Project) did an assessment result in authorities declining to issue required authorizations for projects. In other words, decisions were made in the proverbial "black box," and presumably on the basis of claims about economic benefits that, as Levac, Stinson, and Stienstra point out, tend to be spurious and inequitably distributed.

While, as the authors rightly note, the IAA for the first time imposes a requirement to consider, among other things, the "intersection of sex and gender with other identity factors" (IAA, s 22(1)(s)), there is little reason to hope that this requirement, even alongside the other improvements they list—such as requirements to consider impacts on Indigenous rights, Indigenous and community knowledge, and the extent to which projects will help or hinder Canada's ability to meet its climate commitments (IAA, ss 22(1)(c),(g),(i),(m))—will reverse the trajectory of unsustainable, inequitable, and ultimately politically-driven decisions following assessments.

The authors outline several of the IAA's deficiencies, which impose obstacles to equitable outcomes informed by Indigenous and community knowledge and an intersectional lens. Perhaps the greatest obstacle lies in the nature of decision making itself: At the end of the day, decisions under the IAA remain discretionary, and may be made unilaterally by Crown authorities. The IAA does not require the free, prior, and informed consent of Indigenous Peoples affected by and having inherent jurisdiction over projects subject to assessment, nor does it require decisions to maximize the fair distribution of projects' impacts, benefits, risks, and uncertainties.

The IAA requires the federal environment Minister or Governor in Council (as the case may be) to determine whether the project is in the public interest. This determination is made with regard to the impact assessment report and five enumerated factors: The extent to which the project contributes to sustainability, the significance of federal effects, proposed mitigation measures, impacts on Indigenous Peoples and rights, and the extent to which the project helps or hinders Canada's ability to meet its environmental obligations and climate commitments (IAA, ss 60, 63).

Admittedly, these factors provide more guidance than under previous federal assessment legislation. Additionally, the Minister must issue detailed reasons for a decision (IAA, s 65), which should make it more politically uncomfortable to approve projects with significant greenhouse gas (GHG) emissions or a wildly inequitable distribution of benefits versus climate and other impacts. However, the final decision ultimately remains a political one, in which decision makers may justify the perpetuation of the extractivist mindset that has led us to the climate crisis and continue to advantage those with the greatest economic and social power at the expense of those with the least. In other words, there is no legal bar preventing the Minister from finding that a project that will significantly increase Canada's GHG emissions, increase the risk of violence towards women, and limit Indigenous women's access to important land-based foods and medicines is in the "public interest" because of a few hundred jobs that will largely go to workers from outside the community.

Most critically, the Minister or Governor in Council may make that decision unilaterally, regardless of whether Indigenous Peoples have provided their free, prior, and informed consent or approved the project under their own laws and authority. As Levac, Stinson, and Stienstra point out, inclusive and shared decision making is integral to more equitable and informed decisions that are based on community and Indigenous knowledge alongside western science. Despite multiple calls for the IAA to respect Indigenous laws and decision-making authority, it lacks any requirement to ensure that projects have Indigenous consent, or even to seek consensus with Indigenous Peoples. Notably, the Canadian Province of British Columbia's 2018 *Environmental Assessment Act* (EAA) includes such consensus-seeking requirements (EAA, ss 16(1), 19(1), 27(5), 28(3), 29(3) & (6)(b), 31(5), 32(7)-(8), 34(3), 35(2), 41(5)(c), 41(6), 73(2)). As a result, while decision makers must consider Indigenous and community knowledge, climate change, intersectional impacts and impacts on Indigenous rights, there is no evidence that decisions will be made outside of the black box, or in accordance with the more inclusive and rights-based model that the authors call for.

Two things must change in order for federal impact assessment to hold any promise of meaningfully applying an intersectional lens and, in doing so, countering extractivism and climate change. First, the IAA must be amended to bring it into conformity with the *United Nations Declaration on the Rights of Indigenous Peoples* and in particular, to require decision makers to obtain the free, prior, and informed consent of Indigenous Peoples before making any decisions that affect them or their territories (First Nations Energy and Mining Council, 2021). As Levac, Stinson, and Stienstra note, Indigenous women must be involved in all stages of environmental decision making, and upholding Indigenous rights and authority in impact assessment processes and decisions is a foundational step towards decolonizing environmental planning and management in Canada. Implementing a consent-based approach to impact assessment would also better ensure that Indigenous and community knowledge are more than mere considerations among many others (such as short-term economic benefits accruing to those already holding power and wealth), but instead become the building blocks of informed outcomes that advance equity, sustainability, and climate security.

Second, the IAA should set out a decision-making framework that requires authorities to apply sustainability and equity-based principles and trade-off rules to all decisions (Gibson, 2017; Gibson et al., 2016; Johnston, 2016). Used in a half-dozen EIAs in Canada, including for the Kemess North Copper-Gold Mine, Lower Churchill Hydroelectric Generation Project, and Mackenzie Gas Project, sustainability principles (also sometimes called sustainability criteria) can be used to guide decision makers towards the best option for maximizing environmental, social, economic, and health benefits, minimizing adverse impacts and risks, and helping ensure the equitable distribution of those impacts and benefits. For example, the joint review panel for the Lower Churchill Hydroelectric Generation Project assessment was guided by the following principle: "The effects, risks and uncertainties of the Project should be fairly distributed among affected communities, jurisdictions and generations, and the Project should result in net environmental, social and economic benefits" (Joint Review Panel, 2011 at 269–70). Similarly, the IAA could require that for a project to be found to be in the public interest, its benefits should be maximized and its benefits, impacts, and risks must be as equitably distributed among potentially affected individuals and communities as possible.

Because all resource development decisions entail trade offs, the rationale for trade-off rules is to impose guardrails against unacceptable effects (Gibson et al., 2016), such as those effects that pose considerably more risk to women, Indigenous and racialized Peoples, Peoples with disabilities, and members of 2SLGBTQIA communities. For example, to make the gender-based analysis plus requirement meaningful, the IAA could be amended to prohibit the Minister and Governor in Council from finding that a project is in the public interest if a current or future generation, community, or geographic region will bear an unreasonable share of the adverse risks, effects, or costs, or be denied a reasonable share of the benefits. Additionally, if the IAA prohibited the approval of projects likely to hinder Canada's ability to make good on its climate commitments, it would better ensure that extractivism does not continue to trump the need for immediate and ambitious climate action. Through the imposition of explicit criteria and rules to guide the public interest decision, objectives outlined in this chapter would be much more likely to be actualized in impact assessments, rather than remaining aspirational. The authors' suggested principles and practices are sound. However, there is serious risk that they will not be applied on the ground, as long as doing so remains at the sole discretion of Crown authorities.

## References

*Canadian Environmental Assessment Act*, SC 1992, c 37.
*Canadian Environmental Assessment Act, 2012*, SC 2012, c 19.
*Environmental Assessment Act*, SBC 2018, c 51.

Federal Environmental Assessment Review Office (FEARO). (1987). *The federal environmental assessment and review process*. Minister of Supply and Services Canada.

First Nations Energy and Mining Council. (2021). *Suggested improvements to the Federal Impact Assessment Act by First Nations in BC: Report of the First Nations Energy and Mining Council*. http://fnemc.ca/2021/04/01/iaa-suggested-improvements/

Gibson, R. B. (2017). Applications: From generic criteria to assessments in particular places and cases. In R. B. Gibson (Ed.), *Sustainability assessment: Applications and opportunities* (pp. 16–41). Routledge.

Gibson, R. B., Doelle, M., Sinclair, A. J. (2016). Fulfilling the promise: Basic components of next generation environmental assessment. *Journal of Environmental Law and Practice, 29*, 257–283.

*Impact Assessment Act*, SC 2019, c 28, s 1.

Johnston, A. (2016). Imagining EA 2.0: Outcomes of the 2016 federal environmental assessment reform summit. *Journal of Environmental Law and Practice, 30*, 1–34.

Joint Review Panel. (2011). *Lower Churchill hydroelectric generation project: Report of the joint review panel*. Canadian Environmental Assessment Agency. https://www.ceaa-acee.gc.ca/050/documents/53120/53120E.pdf.

Winfield, M. (2016). Decision-making, governance and sustainability. *Journal of Environmental Law and Practice, 29*, 129–150.

# 4 Tracing Resistance

## Hypermasculinity and Climate Change Denial in the Heart of Alberta's Oil Country

*Angeline Letourneau and Debra Davidson*

***Angeline Letourneau*** *is a third-year Ph.D. student in Environmental Sociology. Angeline has a background in environmental studies and has extensive work experience in both industry and non-profit organizations in Canada and Germany. These experiences have shaped their ongoing research interests in how gendered identities operate in energy and agri-food systems. Their current thesis explores the interplay between resource extraction, gender, and climate change. Angeline's other research interests include theoretical understandings and applications of queer ecology, green criminology, social impact assessments, and the relationship between social media and identity. Their work centres decolonial and intersectional feminist lenses and they centre social justice in all of their work.*

***Debra Davidson*** *is Professor of Environmental Sociology at the University of Alberta, where she has been working since 1999. Her research and teaching are focused on social responses to environmental and climate change, particularly in energy and agri-food systems. She received the Killam Annual Professorship Award (2020), and was a Lead Author on the 5th Assessment Report of the Intergovernmental Panel for Climate Change. Her research is featured in journals such as* Science; Nature Climate Change; Climatic Change; *and* Environmental Research Letters. *She is the co-author of two recent books, including the* Oxford Handbook of Energy and Society *(2018), and* Environment and Society: Concepts and Challenges *(2018). Dr Davidson received her Ph.D. in Sociology from the University of Wisconsin at Madison, in 1998.*

## Introduction

An intersectional, gender-based analysis of climate change calls for an expansive analytical lens, one that encompasses how institutions of gender influence the

DOI: 10.4324/9781003089209-8

production of greenhouse gas emissions, the strategies developed and employed to mitigate those emissions, the distribution of impacts, and the systems of knowledge through which each of these processes is identified and understood. To date, the social science community that is focused on climate change has begun to inquire into this gendered spectrum, with a primary focus on differences in vulnerability and impact between men and women. In the current chapter, we move upstream, as it were, and draw on a range of literature to focus attention to the gendered socio-cultural institutions that support different types of energy systems. We focus on how gendered relations support the fossil fuel economy, and by extension, impose formidable structural resistance to climate change mitigation, with significant political effect.

As we discuss further below, ecofeminists have created important openings into the linkages between patriarchy and environmental destruction (Mies et al., 2014). We begin our inquiry at this linkage, but then dig deeper into how gendered relations support the fossil fuel economy, even in the face of evidence that it is irrefutably damaging to the environment and the climate. Although patriarchy is sustained through several mechanisms, one of these is the endorsement of certain expressions of masculinity. While expressions of masculinity, like all manifestations of identity, are dynamic and diverse, different cultural contexts can give rise to masculinities that may be considered "hypermasculine" or even toxic (Russell, 2021). The degree to which any expression of masculinity falls into either of these categories or, in contrast, is celebrated and held in high regard, is heavily dependent on the specific culture from which it is viewed.

As we will illustrate, the celebration of specific expressions of masculinity can become toxic. We apply a critical feminist lens to fossil fuel extraction zones, places where hypermasculinity has been allowed to flourish with violent results for women and ecosystems, but also with substantial implications for climate change mitigation and adaptation. We hope to highlight that, so long as toxic forms of masculinity maintain a strong foothold in the cultural psyche, fossil fuels will continue to sustain energy dominance and social acceptability.

## Background

Incorporating a feminist critical analysis into studies of environment-society relations is not novel. Feminist scholars have, for example, offered insightful critiques of the production and application of western science, including in the realm of environmental management (i.e., Haraway, 1988; Harding, 1991). Relatedly, ecofeminists have offered a unique critique of both human domination and the domination of nature, extending beyond gender-based discrimination to encompass oppression along racial and class divides and how these human categories have been associated with the natural world to justify their continued domination. Many ecofeminist theorists have provided powerful critiques of modern western culture, capitalism, colonialism, and androcentric epistemologies (Cudworth, 2005; Gaard, 2001; Haraway, 1991, 2016; Mies et al., 2014). There has also been substantial work on the intersectional impacts of climate change for women (e.g., Kaijser & Kronsell, 2014; MacGregor, 2009; Terry, 2009), many of which are highlighted in other chapters of this volume.

In this chapter, we will draw from an emerging field of scholarship on gender dimensions of energy systems. To begin, some empirical researchers have drawn attention to women's limited access to energy (Johnson et al., 2019; Oparaocha & Dutta, 2011; Pueyo & Maestre, 2019), and their underrepresentation in the energy workforce and decision making (Allison et al., 2019; Fraune, 2015). A small but highly relevant set of papers has explicitly applied a feminist studies lens to energy systems (e.g., Bell, 2016; Bell et al., 2020; Daggett, 2019; Hultman, 2013; Ryan, 2014; Ryder, 2018; Wilson, 2014), including the intersections between masculinity and energy production (Bell, 2016; Bell & Braun, 2010; Bell & York, 2010; Daggett, 2018, 2019; Hultman & Anshelm, 2017; Scott, 2010). These scholars draw connections between how masculinity as an identity is often intimately tied to occupational engagement in specific forms of energy development, such that threats to those industries may be perceived as threats to personal identity.

We seek to integrate and expand upon existing studies of gender and energy with a particular focus on masculinity and fossil fuel extractive economies. We argue that fossil fuel-based economic sectors, which pose among the most formidable of political constraints to effective climate change mitigation, are deeply entangled with masculinity. Confrontations with fossil fuel interests in climate politics thus entail confrontations with these structures of masculinity, and resistance to the science and politics of climate mitigation may be viewed as a form of defence of masculine identities. In the section below, we draw from masculinity studies to postulate the interlinkages between masculinity and fossil fuel extraction. Following this, we explore how such interlinkages manifest in Alberta's extraction zones. In closing, we discuss the implications for climate change mitigation, impacts, and adaptation in a social context that prioritizes petro-masculine ideals and call for a reimagining of masculinity and the gendered structures that situate masculinity and femininity in opposition to one another.

### *Masculinity Literature*

What do we mean by masculinity, and how does it have bearing on energy and climate systems? Masculinity studies define a broad field from which we draw to make our case, highlighting several key dimensions: The constructed nature of masculinity; masculinity and hegemony; toxic masculinity; linkages between masculinity and environmental concern; and "petro-masculinities," an emerging conceptual lens introduced by Cara Daggett (2018, 2019).

### *What Is Masculinity?*

Masculinity refers to a set of culturally defined attributes and practices that come to be associated with and expected of men, often as a means of differentiating them from women (Messerschmidt, 2018). In patriarchal societies masculinity is associated with power; however, it is crucial to differentiate between patriarchy, as a system of oppression against women, and gender, which describes social constructions of identity, including multiple forms of masculinity.

Masculinity is an expression of identity, and as such, it is socially constructed, emerging in daily practices and relationships. This means there are multiple representations of masculinity, and they are not static. Like most identifiers, masculinities are constructed and navigated through everyday interactions and practices, existing beyond any given individual (Connell, 1995; Holter, 1997). Because masculinities must be performed as a social action, they differ greatly depending on the gender relations of a particular social context (Connell & Messerschmidt, 2005). In this way, masculinities do not exist in isolation from social behaviour, rather, they come into existence as people navigate different social spaces (Connell, 2001). They are accomplished in everyday conduct or organizational life, as patterns of social practice. Given the importance of social context in defining and characterizing masculinities, the relevance of identifiers such as geographic region and occupation becomes clearer, as they play a significant role in shaping the given social context that may influence the expression of masculinity.

*Masculinity Can Be Expressed Hegemonically*

The multiple forms of masculinity are not equally or consistently valued across different contexts. Connell (1990, p. 454) conceptualizes masculinity as a socially constructed identity that emerges "within a gender order that defines masculinity in opposition to femininity, and in so doing, sustains a power relation between men and women as groups." This creates a need to differentiate or distance one's masculinity from femininity, thus creating a hierarchy of masculinities with those most distinct from femininity at the top, and those with more ambiguity lower. Preference for particular masculinities in different social contexts, contrasted with the mere tolerance of or outright rejection of others, reveals hierarchies of masculinity. The form of masculinity that is dominant in any given social context is "hegemonic masculinity" (Connell, 2001). The hegemonic form of masculinity is not necessarily the most common. Instead, hegemonic masculinity represents a prioritized or elevated status in gender politics overall, illustrating the privilege that men—particularly those who embody hegemonic masculinity—have over women. In other words, the hierarchy of masculinities reflects the unequal dispersion of that privilege among different groups of men (Connell, 2001).

While by no means a rule, the manifestation of hegemonic masculinity can be associated with the use or threat of violence as a means of defending hegemonic status, termed toxic masculinity by some researchers, which we discuss further below (Connell & Messerschmidt, 2005). It is the all-too-well-known toxic practices of hegemonic masculinity that inspire Donaldson's (1993) description of hegemonic masculinity in western culture:

[As] a culturally idealized form, it is both personal and a collective project and is the common sense about breadwinning and manhood. It is exclusive, anxiety-provoking, internally and hierarchically differentiated, brutal, and violent ... constructed through difficult negotiation over a lifetime. Fragile it may be, but it constructs the most dangerous things we live with.

(Donaldson, 1993, p. 645–6)

Donaldson's characterization of western hegemonic masculinity is instrumental in allowing us to consider the various regional influences on exactly how western hegemonic masculinity presents itself. While the general themes that Donaldson articulates may very well be accurate, further details would begin to become regionally specific. For example, while Vancouver, Edmonton, and Ottawa are all part of Canada, and therefore similar in many of their cultural elements, one would still not expect the hegemonic masculine in any of these cities to be the same. They would certainly be similar in a broader sense, as Donaldson describes, and may have more in common with each other than with the hegemonic masculinity describing a set of rural villages in China, but what may be celebrated as the most acceptable way to be a breadwinner or express "manhood" will likely be different in any of these cities based on their vastly different social contexts. On the other hand, "cultural consent, discursive centrality, institutionalization, and the marginalization or delegitimation of alternatives are widely documented features of socially dominant masculinities" that one would expect to see across all three locations (Connell & Messerschmidt, 2005, p. 846).

## Toxic Masculinity

Kimmel (2017) further noted that not all men are in positions of power, and many do not necessarily feel powerful at all. Men as a group hold power over women as a group; however, some men also hold power over other men due to the intersections of class, race, ethnicity, and other systems of power. This can inform a sense among men, particularly White men, that they *should* feel powerful—are entitled to feel powerful—and when they do not, violence is often a means through which individuals seek to restore power and respect (Bundy, 2003, as cited in Kimmel, 2017).

As masculine identities come under threat, the potential for such malignant tendencies increases significantly, manifesting in what has been referred to as toxic masculinity. Daddow and Hertner (2019; see also Jenney & Exner-Cortens, 2018; Schippers, 2007) locate toxic masculinity at the most extreme and violent end of hegemonic masculinity. Some of the more familiar characteristics of toxic masculinity include rage (Haider, 2016), the tendency to suppress empathy for others (Romero, 2017), and inadequate communication styles in person-to-person and online discourses (Parent et al., 2018). Of course, not every male-identifying person will exhibit these traits, much in the same way that not every male-identifying person likes cars. Certain men more than others will feel compelled to emulate a hegemonic form of masculinity, and these individuals will perceive threats to their efforts to do so, and respond to those threats, in different ways. Understanding both the ways that toxic masculinity manifests and how a perceived threat to hegemony may be responsible for such toxic outbursts is important in understanding the connections between masculinity and the fossil fuel industry.

## *Masculinity and the Environment*

As noted by ecofeminists, patriarchy equates and denigrates both women and the natural world. Griffin (1997, p. 219) argues that the "social construction (exploitation, destruction) of nature is implicit in and inseparable from the social construction of gender." What this looks like on the ground in certain cultural contexts involves, first, the feminization of environmental concern, which can become particularly blatant in a rural, extractive landscape. Expressions of concern for the environment, and the individuals expressing such concern—whether urban-based environmentalists or university-based climate scientists—are "feminized" and thus to be opposed by individuals who embrace the masculine identity of that extractive culture (Anshelm & Hultman, 2014). This is particularly true for masculine identities that align with conservative political ideologies, as they have been increasingly linked to climate skepticism and resistance to environmental regulation. A significant body of empirical research has consistently identified strong associations between climate skepticism and conservative White male identity (e.g., McCright & Dunlap, 2011).

Second, when environmental concern is accommodated, such issues are approached from a lens of control over and utilization of the natural world, through the masculinized tools of science and technology. This lens is most readily observed in ecomodernism, whose most prominent flag bearers are elite, White male northerners such as Bjorn Lomborg, Michael Schellenberger, and Elon Musk (Daggett, 2018). Ecomodernists claim that there is no inherent conflict between the environment and the economy and that technological solutions generated through the free-market system can solve any challenges that may arise (Hultman, 2013). Hultman traces the rising acceptance of ecomodernism with a shift in hegemonic masculinity through the career of Arnold Schwarzenegger, former Governor of California. More traditional, industrial-era masculinities were embodied by Schwarzenegger in his younger days of powerlifting and machismo, and these were traded in for the hybrid masculinity of Governor Schwarzenegger of California, champion of green technologies and driver of the greenwashed Hummer (see also Hultman, 2013).

The new masculinity created by ecomodernists like Governor Schwarzenegger created space for compassion and care to coexist alongside traditionally masculine traits like toughness (Daggett, 2018); however, this compassion remained secondary to techno-rationality and economic growth. While the door was opened to consider more "feminine" traits, the decision-making process remained asymmetrical in priorities (Hultman, 2013).

*Masculinity and Fossil Fuels*

Daggett (2018) digs deeper into the empirical associations between gender and climate skepticism, postulating that the relationships between authoritarian movements, climate change denial, racism, and misogyny can be understood by applying the concept of "petro-masculinity." Building on the work of Connell, Daggett situates petro-masculinity within a gender order that perceives femininity and masculinity to be in direct conflict. In contrast to environmental masculinities, such as ecomodernism, petro-masculinity does not attempt to appear concerned about the wellbeing of other people or the planet. Drawing upon aspects of traditional hegemonic masculinity, petro-masculinity is best understood as a type of hypermasculinity wherein actors lash out when they feel their power is threatened. In contrast to environmental masculinities—namely ecomodernism—petro-masculinity does not seek to obtain hegemony in some new technoscientific utopia; rather, it aims to defend petro-cultures from those who may seek to undermine their absolute power, namely, climate advocates. This petro-masculinity may be a key source of defence of the sector's status as "potent conservative symbols that represent autonomy and self-sufficiency" (Daggett, 2018, p. 35), even when confronted by the growing market vulnerability.

Termed "fossil violence" by Daggett, the intersectional impacts of extraction activities are extensive; however, instead of ignoring climate change, racism, or gender inequality and remaining inactive, petro-masculinity is reactionary and actively denies these realities. For some, this denial may be a welcome reprieve from the guilt, uncertainty, and despair that usually follow in the face of conversations on global warming. By simply narrowing the scope of beings that "matter" to White men and prioritizing their comfort and wellbeing alone, the issues caused by fossil violence cease to be problems at all.

The dismissal of these intersectional impacts of fossil violence become particularly vivid and hostile in the realm of environmental politics, within which climate action has moved to the epicentre. Many studies have commented upon the high levels of representation of women in environmental activism (Brown & Ferguson, 1995; Shriver et al., 2013), and women's involvement appears to be even more prominent in local environmental justice mobilizations. Many women become involved in environmental justice movements because they are concerned for their children's or grandchildren's wellbeing (Bell & Braun, 2010; Brown & Ferguson, 1995; Culley & Angelique, 2003; Krauss, 1993; Peeples & Deluca, 2006). In contrast, men may be even less likely to engage in such actions because of locally specific masculine identities as workers and economic providers in the industries targeted (Bell & Braun, 2010; Bell & York, 2010).

## Masculine Petro-Cultures at Work in Alberta

Since the drilling of the first oil well in the Western Canadian province of Alberta in 1902, the settler culture and economy has been defined by deep entanglements between ranching and oil production, generating a stable provincial identity that persists to this day. Part of that stability can be attributed to the highly complementary—and deeply gendered—character of both industries. Each is founded upon similar cultural themes of frontier, individualism, and risk-taking, all of which are embodied in the romantic image of the cowboy hero, equally applicable to ranchers and oilmen (indeed, they were often one and the same) (Miller, 2004; Wright, 2001).

In Alberta today, much would appear to have been untouched by time, with the same iconic images of landscapes depicting stark, rugged mountains behind expansive ranches, fields of grain, and the ever-present oil rig. Centre stage in these images are the same occupational players of yesteryear: The stoic rancher, grain farmer, and rig worker, clad in tough canvas or leather work pants, muddy, steel-toed boots, and always a hat—perhaps a wide-brimmed cowboy hat, trucker's cap, or bright yellow hard hat.

However, these images belie significant transformations in these industries. Most Albertans live and work in the province's urban centres now, and the proportion of jobs supported by energy and agriculture has declined precipitously. According to Natural Resources Canada (2020), as of 2019, the national energy sector provided direct employment for 282,000 people in Canada, a fraction of a national workforce that includes over 18 million. About half of these jobs are in Alberta, making up approximately 6% of the provincial workforce; a proportion that has been steadily declining for over a decade (Alberta Labour and Immigration, 2020). The scale of development, and the size of associated job-replacing machinery, have escalated in tandem, including giant trucks, loaders, and drills, which explains why the sector accounted for 27% of provincial GDP in 2018, despite the shrinking labour force.

While the proportional number of jobs in the oil sector has declined, the frontier cowboy mythology lives on: Success follows hard work, thick skin, and an entrepreneurial spirit. Indeed, over much of the first two decades of the 21st century, thanks to record-high oil prices and technological developments that enabled the commercialization of the enormous reserves of bitumen that lay under the northeastern corner of the province, this dream has closely resembled reality for many able-bodied men who entered the oil and gas sector. From highly skilled trades individuals to inexperienced laborers, the wages for any job in the energy sector in Alberta are inflated well beyond their counterparts in any other sector (Statistics Canada, 2020b). Thanks to educational policies from premier Ralph Klein throughout the 1990s and early 2000s, students have been able to begin their trade apprenticeship programmes while still in high school, potentially earning the same salary as their high school teachers within a few years of graduating (Government of Alberta, 2020b). Notably, oil and gas sector jobs are accessible to individuals from lower income backgrounds, with apprenticeship programmes only requiring a total of eight months of formal education, spread out over four years, eventually providing access to the six-figure income of a journeyman tradesperson.

Work environments have remained highly male-dominated, despite industry-backed programmes to encourage more women to pursue occupations in the industry. Just 22% of industry workers were women in 2019, and over 90% of higher-waged tradespeople and equipment operators were male (Energy Safety Canada, 2021). Instead, women employed by oil and gas are significantly more likely to work in the lower-paid, feminized spaces within company offices, overseeing administrative tasks (Dorow & Mandizadza, 2018).

There are numerous reasons for this gender gap, including the structure of these occupations. Residing away from home for weeks at a time is certainly not compatible with childcare responsibilities, for example (Dorow & Mandizadza, 2018). Perhaps another reason is the highly masculinized work culture associated with these work environments. The imagery provided to would-be entrants into the field has changed little over the decades, as depicted in the two cover photos of the industry magazine *Oilweek* below. Published nearly three decades apart, both images portray oil workers as rugged, risk-taking individuals battling the elements and applying brute strength to wrest resources from the frontier.

Less readily observable, but no less consequential, are the gendered social-interactional practices that resonate in this sector. Miller (2004) describes these practices in detail, from the drill rig to the boardroom. Female workers must learn to navigate an environment that includes exclusive male-dominated informal networks as spaces through which upward career mobility must pass (envision, for example, the proverbial golf outing); the pre-eminence of confrontational, "tough" negotiation tactics; and idealization of risk-taking as the route to success (Miller, 2004).

Perhaps the most tangible embodiment of this celebrated hypermasculinity is the so-called "man-camp." Situated on the geographic fringes of the province, surrounded by fences and often served by little more than a hard-packed dirt road,

*Figure 4.1* Magazine covers from *Oilweek* magazine, 1978 and 2004. *Source:* Canadian Energy Museum

man-camps have been instrumental in allowing for remote resource extraction in regions with little or no infrastructure. These camps provide lodging, sewage and water treatment, police, and social support services to accommodate sharp increases in temporary populations. The term "man-camp" is reflective of the individuals these camps are designed to serve—predominantly White, male extraction workers—rather than the relatively fewer racialized or female bodies employed to fulfil the social reproductive roles necessary to keep the camp functioning (Dorow & Mandizadza, 2018). The men occupying the camps all perform a similar role: They work long, physically demanding hours out-of-doors, in work boots and hard hats, moving from site to site, to wrest fossil fuels from beneath the earth.

Given this stark homogeneity, man-camps serve as a petri dish for cultivation of a hegemonic form of hypermasculinity, where a rigid and narrow spectrum of masculine ideals is strongly enforced, and the constant rotation of new workers requires the continued need to prove oneself. Regardless of their origins, workers become completely immersed within the social norms created by this homogenous, hypermasculine social environment (Dorow & Mandizadza, 2018).

One outcome of this socio-environmental setting is an elevated potential for engagement in dysfunctional behaviors on the part of stressed, overworked men who struggle to meet the expectations associated with this hegemonic masculinity. The isolation and hypermasculinity that define day-to-day life for extraction workers place significant stress on individuals with limited options for a healthy outlet (Dorow, 2015). Distress, loneliness, physical injury, even grief over the destruction taking place around them, could all be interpreted as potential signs of "weakness" that must be concealed in such a hypermasculinized social field. The fossil fuel model of temporary, mobile labour facilitates such behaviors. Thus, it may not be surprising to observe problematic after-hours behaviours, which workers may feel are justified, as a form of "blowing off steam"—for example, partaking in the nightlife offered in nearby rural communities. Given the temporary nature of their lives within remote, rural regions, fossil fuel workers are often disconnected from those communities, and, likewise, the social norms that may otherwise sanction toxic behaviours.

Oil and gas industry workers have significantly higher rates of alcohol use than the average for any other industry (Alberta Alcohol and Drug Abuse Commission, 2006; Alberta Health Services, 2017). While it is difficult to gain a holistic perspective of the extent of the substance abuse problems in extractive sectors, companies such as the oil sands giant Suncor have faced extensive issues with employee and contractor substance abuse (Thurton, 2017). While it may be tempting to conclude that excessive substance abuse is just a reality for anyone working in extractive industries that require workers to be away from their families and communities for so long under grueling working conditions, gender still appears to play a role: Male oil and gas workers are more likely to be classified as medium or high risk for experiencing problematic drinking, while women in the oil and gas industry are more likely to be classified as low risk (Alberta Health Services, 2017).

Additionally, and most grievously, these dysfunctional behaviors—intersecting with a hypermasculinized culture that, as highlighted earlier, must alienate itself from femininity to gain hegemony—can extend toward the treatment of women. Indigenous women in particular have been targets of such dysfunctional behaviours. Indigenous women across Canada are at a greater risk of violence than non-Indigenous women: The 2014 General Social Survey on Victimization found that Indigenous women were six times more likely to be victims of homicide and three times more likely to be victims of domestic violence than non-Indigenous women (Statistics Canada, 2014; see also National Inquiry into Missing and Murdered Indigenous Women, 2019). Rates of violence against Indigenous women appear particularly high in remote communities with a high extractive industry presence and associated presence of transient, male workers (Jones & Fionda, 2017). The National Inquiry on Missing and Murdered Indigenous Women's report released in June 2019 explicitly warned of the observable increases in crime levels, sexual offences, and domestic and "gang" violence in remote resource development contexts.

### *A Provincial Identity*

The hegemonic hypermasculinity flourishing in the oilfields of Alberta extends beyond those fields and associated executive boardrooms. This cultural fingerprint can be clearly seen throughout the province's populist and conservative political realm, in which the "right" to exploit "our" resources is fiercely defended, implicitly or explicitly in the name of the hypermasculine, natural resource-hewing male. Oil workers, and the politicians who seek their support, express a sense of entitlement to their continued petro-culture and its accoutrements: The readily available high-income jobs, the resulting high tax revenue that allows Alberta to remain the only province in Canada without a provincial sales tax, and an extreme form of neo-liberal, trickle-down economics that embraces that century-old romantic image of the do-it-yourself cowboy. Even Rachel Notley, leader of the left-leaning New Democratic Party, which controlled the Legislature from 2015 to 2019, perceived the need to make public appearances in her cowboy hat to affirm her identity as a true Albertan.

### *Troubling Times Ahead*

That sense of entitlement to prosperity is facing two formidable and inter-related threats. First, the historically "boom and bust" oil and gas industries appear to be on a precipitous economic downturn that began with declining oil prices in 2014 (Statistics Canada, 2020a). Jobs in oil and gas extraction, already facing a continuous decline due to the increasing mechanization and automation of work, were shed even faster.

Second, climate change has slowly but steadily risen of to the top of national and international political agendas, placing Alberta in a particularly tarnished light, and creating a significant legitimacy challenge for the federal government, which committed to significant reductions in national greenhouse gas emissions by sign-ing the Paris Agreement. Alberta may be primarily responsible for what increas-ingly appears to be an inevitable failure to meet those commitments; in 2019, Alberta was Canada's highest emitter of greenhouse gas emissions (Environment and Climate Change Canada, 2021). As with all political arenas, climate poli-tics has already spilled over into our economies, and analysts are increasingly referring to oil and gas reserves as *stranded assets*, potentially losing their value due to the climate mitigation imperative to eliminate emissions. Investors have responded by selling off their shares in Alberta's energy sector (Van der Ploeg & Rezai, 2020).

Sensitive to this changing international landscape, former Premier Notley sought to accommodate these political pressures with a new provincial carbon tax, among other initiatives. However, upon election in 2019, one of the first moves of Alberta United Conservative Party (UCP) leader Jason Kenney—a passionate petro-fundamentalist—was to fulfil his promise to repeal the carbon tax, while also launching a campaign to discredit climate activism.

Premier Kenney tweeted the following response to remarks made by Prime Minister Justin Trudeau in a recent speech (Dawson, 2018) regarding newly insti-tuted federal environmental impact assessment requirements:

> Justin Trudeau says pipelines must go through a 'gender-based analysis' because male construction workers have 'impacts.' Darned right they do. They build things, create wealth, pay taxes, take care of their families. But this trust fund millionaire thinks they can't be trusted. (Kenney, 2018)

While standing up for jobs in the oil and gas sector amid a downturn in the global oil market, elected officials in Alberta have simultaneously pursued steep public sector cuts to education and health care—sectors that also happen to employ a disproportionate number of women in a province where women are already dis-enfranchised in the labour market. According to the Women's Centre of Calgary (WCoC), "women in Alberta face the largest unemployment gender gap of any province, are overrepresented in lower-paying and minimum-wage jobs, and experience poverty at a greater rate than men" (WCoC, 2020).

The energy industry jobs touted by the Premier are not explicitly exclusive to men; however, such jobs may exclude women in other ways. One of the barriers to women entering the higher paid extractive industry sector is the unpredictable schedules and long hours that define most of this work. This job structure signifi-cantly limits those who can take on such high-paying roles to individuals who have very few personal commitments—such as younger, single individuals—or to those who have a caretaker at home to provide social reproduction labour like childcare (Dorow, 2015). While hired childcare is certainly an option, the gen-dered social norms in this extractive culture punish women who would choose to leave their children for any period to work, pressuring women to avoid the well-paying industries in extraction, or at the very least, to leave the sector after having children (Dorow, 2015; Dorow & Mandizadza, 2018). Social norms thus place a great deal of pressure on women with career ambitions to work the double shift and manage both their careers and the reproductive care of their families. This same pressure does not exist for men, and indeed, men in the fossil fuel industry are often praised for sacrificing their time with their family to send home a pay-check—as evidenced by Kenney's tweet.

The strong defence of the extractive industry by elected officials can be seen as the political corollary to the hypermasculinity of the man-camp, as Alberta's political defenders take a tough stance against any "outsiders" seeking to attack Alberta. Drawing a figurative steel-toed line in the sand, anyone—inside or outside the province—who expresses concern for climate change and raises the potential need for downscaling fossil fuel production, is labelled as "anti-Albertan." The UCP has gone so far as to fund a public inquiry into climate advocacy, depicted as a foreign funded "anti-Albertan" conspiracy.[1]

Such "anti-Albertans" may indeed pose threats to Alberta's petro-economy by exposing the ruling UCP's campaign to cut "red tape"—particularly red tape that takes the form of environmental regulation and climate mitigation policy. The provincial government justifies such actions with a discourse of efficiency, citing the imperative to "save time, create jobs, or reduce costs," particularly during the pandemic (Government of Alberta, 2020a). The Canadian Association of Petroleum Producers (CAPP), a national lobby group representing much of the fossil fuel sector, could not agree more. CAPP has certainly done their part to defend the frontier myth of Alberta, compatible as it is with the profits of oil and gas interests, not only through lobbying, but also with carefully orchestrated social media campaigns. For hypermasculine defenders of Alberta's oil and gas sector, advocates for climate mitigation are feminized and subjected to a discourse of sexualized violence. This violent manifestation of toxic masculinity was put on vivid display in February 2020, when a meme was circulated by an oil and gas sector company showing climate youth activist Greta Thunberg being sexually violated. Other common targets of such threats included the Honourable Catherine McKenna, the former federal Minister of Environment and Climate Change, and former Provincial Minister of Environment Shannon Phillips. The most recent confrontation emerged in the form of a report by the International Energy Agency (IEA) that landed like a lead balloon in the boardrooms of oil, gas, and coal companies the world over, proclaiming in unambiguous terms that their proposed pathway to net zero (greenhouse gas emissions) by 2050 entails *no new investments in fossil fuel supply* (IEA, 2021).

## Implications of Petro-Masculinity

What implications does petro-masculinity have for provincial politics in Alberta, and the world more broadly? Climate denial in Alberta goes much deeper than a political-economic phenomenon—it represents the defence of a threatened hyper-masculine identity. Climate change mitigation strategies pose significant threats to male-dominated oil-worker jobs and lifestyles. These threats extend beyond the blue-collar worker—the image that is typically invoked when the topic of labour in the energy industry is raised—and extend up through the white-collar workforce before diffusing throughout the political elite in a challenge against the provincial self-image. Those who embody this threat—climate scientists and environmental advocates—become feminized in a narrative that attempts to re-establish a hypermasculine hegemony.

Hypermasculinity offers a formidable source of underlying support to petroleum industries. Threats to petroleum industries are thus received as threats to petro-masculinity and responded to in a requisite manner. Defenders lash out against perceived enemies, in defence of the industry but also of their masculine identities, in part by feminizing those perceived enemies to re-establish a sense of control. As part of this cultural battle, petroleum industry advocates have become staunchly opposed to environmental and climate policies while continually reaffirming their support for fossil fuels—even in the face of the growing reality that to do so is acutely harmful to the environment, the climate, and increasingly, even to the regional economy. Any policies restricting extraction activities are dismissed as costly luxuries, and environment and climate advocates, many of whom are women, are likewise dismissed as "enemies" of Alberta.

Gender roles in western cultures adhere to a rigid gender binary (e.g., Connell, 1990; Moskos, 2020), with masculinity and femininity positioned as opposites. Patriarchy operates from within this binary to prioritize masculinity and devalue practices, views, and identities that are associated with the "feminine," thereby ensuring continued power and control. Consequently, while "feminine" elements may be given a slight nod by some masculinities (such as ecomodernism), these elements will remain secondary to the more familiar masculine ideals like meritocracy, technology, and control. Because western patriarchy—the system that produced petro-masculinity—presumes unavoidable friction between masculinity and femininity, it establishes a scenario where a feminized nature must also conflict with masculinity (e.g., Mies et al., 2014; Plumwood, 2018). In this space, bowing to outside pressure and conceding to (feminized) climate scientists and advocates would constitute a loss of autonomy and control, thus undermining the respect that provincial leaders have from citizens. These socially constructed, gendered dichotomies silence constructive debates and instead present to Albertans (and Canadians) the fallacious choice of supporting the climate or supporting their province/country; a future of fossil fuel derived prosperity or a future of disempowerment and poverty.

## Conclusion

In a social and political landscape where oil and gas extraction has produced significant wealth but is now under increasing pressure to address environmental and climate impacts, the defensive posturing of Alberta's hypermasculine identity is on full display. Through its celebration of the industrial, self-sacrificing working man, hypermasculinity acts as a proud rallying cry for the male breadwinner who wrests raw resources from the earth. This identity has become sufficiently hegemonic to be idealized well beyond the boundaries of man-camps and industrial facilities. Instead, this identity is embraced broadly within Albertan culture, to the extent that expressing at least some support for it becomes necessary for aspiring political leaders to gain electoral support.

Ensuring the continued hegemonic petro-masculinity within Alberta today, given the environmental and economic threats the industry faces, requires a significant amount of work. Although political discourse proclaiming protection of Alberta workers implies gender neutrality, substantial government support of the oil and gas industry and its male-dominated workforce during the current period of economic decline has coincided with catastrophic cuts to education, healthcare, and other public sectors which have all had a disproportionate effect on working women.

The petro-masculine ideal of hard-working, strong, and independent male workers belies the enormous challenges that extraction workers face, exposed to all manner of elements in full PPE (personal protective equipment) and maintaining demanding work schedules that often take them away from their families for weeks at a time. Workers are expected to "man up" and not complain. The inability to be transparent about the obvious challenges that they face being in such unforgiving environments creates the necessary conditions for a host of toxic outcomes, such as drug and alcohol abuse, domestic abuse, and rape culture.

As pressure grows to mitigate climate change, animosity and violence have become directed toward climate change advocates such as Greta Thunberg. The outcome is an explosive situation regionally, with global repercussions as these extractive zones become powerful nodes of resistance in international efforts to combat climate change, manifesting in strong opposition to even modest proposals for mitigative strategies, such as carbon taxes.

Progress in climate mitigation requires confronting and dismantling the gendered structures that have served to buttress the fossil fuel industry. This process begins with the acknowledgement that petro-masculinity is a deeply entrenched social identity in Alberta and, as with all identities, will not be readily shed by its adherents. Given the hegemony of this identity in extractive cultures, to relinquish it would be to relinquish political power. The second step forward entails a deeper questioning of certain truisms in western culture that prioritize economic growth over human and ecological wellbeing, and idealization of masculine over feminine traits.

As unsustainable as the hegemonic masculinities in western extractive culture may be, other forms of masculinities exist. Ecological masculinity is one such example. Valuing practices such as the localization of economies, and the creation of small-scale technologies and decentralized renewable energy systems, ecological masculinity prioritizes cohabitation with nature in everyday life (Hultman & Anshlem, 2017). We hope that the embracing of ecological masculinity by actors in groups such as 350.org and MenEngage serves as inspiration for how masculinities may be reimagined and reclaimed to align with the pressing need for action.

This renegotiation of masculine identities is not just about the need for action on climate goals—though this may be the most pressing application of it—but it highlights the need for a new hegemonic masculinity: One which moves beyond the patriarchal binary of men and women and, instead, engages with ecological values and empathetic practices.

## Note

1  For more information, see https://albertainquiry.ca

## References

Alberta Alcohol and Drug Abuse Commission. (2006). *Substance use and gambling in the workplace, 2002: A replication study*. http://www.assembly.ab.ca/lao/library/egovdocs/2003/alad/144909.pdf

Alberta Health Services. (2017). *Alcohol and drug use in Alberta's oil and gas industry*. https://www.albertahealthservices.ca/assets/info/res/mhr/if-res-mhr-alcohol-drug-oil-gas-industry.pdf

Alberta Labour and Immigration. (2020). *Alberta mining and oil and gas extraction industry profile, 2018 and 2019*. https://open.alberta.ca/publications/alberta-mining-and-oil-gas-extraction-industry-profile

Allison, J. E., McCrory, K., & Oxnevad, I. (2019). Closing the renewable energy gender gap in the United States and Canada: The role of women's professional networking. *Energy Research and Social Sciences*, *55*, 35–45. https://doi.org/10.1016/j.erss.2019.03.011

Anshelm, J., & Hultman, M. (2014). A green fatwā? Climate change as a threat to the masculinity of industrial modernity. *Norma*, *9*(2), 84–96. https://doi.org/10.1080/18902138.2014.908627

Bell, S. E. (2016). *Fighting king coal: The challenges to micromobilization in Central Appalachia*. The MIT Press.

Bell, S. E., & Braun, Y. A. (2010). Coal, identity, and the gendering of environmental justice activism in Central Appalachia. *Gender & Society*, *24*(6), 717–745. https://doi.org/10.1177/0891243210387277

Bell, S. E., & York, R. (2010). Community economic identity: The coal industry and ideology construction in West Virginia. *Rural Sociology*, *75*(1), 111–143. https://doi.org/10.1111/j.1549-0831.2009.00004.x

Bell, S. E., Daggett, C., & Labuski, C. (2020). Toward feminist energy systems: Why adding women and solar panels is not enough. *Energy Research & Social Science*, *68*, 101557. https://doi.org/10.1016/j.erss.2020.101557

Brown, P., & Ferguson, F. I. T. (1995). 'Making a big stink': Women's work, women's relationships, and toxic waste activism. *Gender & Society*, *9*, 145–172. http://www.jstor.org/stable/189869

Connell, R. (1990). A whole new world: Remaking masculinity in the context of the environmental movement. *Gender and Society*, *4*(4), 452–478. https://doi.org/10.1177/089124390004004003

Connell, R. W. (1995). *Masculinities*. Berkeley.

Connell, R. W. (2001). Understanding men: Gender sociology and the new international research on masculinities. *Social Thought & Research*, 24(1), 13–31. https://www.jstor.org/stable/23250072

Connell, R. W., & Messerschmidt, J. W. (2005). Hegemonic masculinity: Rethinking the concept. *Gender & Society*, *19*(6), 829–859.

Cudworth, E. (2005). Introduction. In E. Cudworth (Ed.), *Developing ecofeminist theory: The complexity of difference*. Palgrave Macmillan.

Culley, M. R., & Angelique, H. L. (2003). Women's gendered experiences as long-term three mile island activists. *Gender & Society*, *17*(3), 445–461. https://doi.org/10.1177/0891243203017003009

Daddow, O., & Hertner, I. (2019). Interpreting toxic masculinity in political parties: A framework for analysis. *Party Politics*, *27*(4), 743–754. https://doi.org/10.1177/1354068819887591

Daggett, C. (2018). Petro-masculinity: Fossil fuels and authoritarian desire. *Millennium: Journal of International Studies*, *47*(1), 25–44. https://doi.org/10.1177/0305829818775817

Daggett, C. (2019). *The birth of energy: Fossil fuels, thermodynamics, and the politics of work*. Duke University Press.

Dawson, T. (2018, December 4). What you need to know about Trudeau's comments on 'gender impacts' and construction workers. *National Post*. https://nationalpost.com/news/canada/heres-what-you-need-to-know-about-the-controversy-over-trudeaus-comments-about-gender-impacts-and-construction-workers

Donaldson, M. (1993). What is hegemonic masculinity? *Theory and Society*, *22*(5), 643–657. https://doi.org/10.1007/BF00993540

Dorow, S. (2015). Gendering energy extraction in Fort Mcmurray. In M. Shrivastava & L. Stefanick (Eds.), *Alberta oil and the decline of democracy in Canada*. AU Press.

Dorow, S., & Mandizadza, S. (2018). Gendered circuits of care in the mobility regime of Alberta's oil sands. *Gender, Place & Culture*, *25*(8), 1241–1256. https://doi.org/10.1080/0966369X.2018.1425287

Energy Safety Canada. (2021). *Employment and labour data*. https://careersinenergy.ca/employment-and-labour-data/

Environment and Climate Change Canada. (2021). *Greenhouse gas emissions: Canadian environmental sustainability indicators*. Government of Canada. https://www.canada.ca/en/environment-climate-change/services/environmental-indicators/greenhouse-gas-emissions.html

Fraune, C. (2015). Gender matters: Women, renewable energy, and citizen participation in Germany. *Energy Resources and Social Sciences*, *7*, 55–65. https://doi.org/10.1016/j.erss.2015.02.005

Gaard, G. (2001). Women, water, energy: An ecofeminist approach. *Organization & Environment*, *14*(2), 157–172. https://www.jstor.org/stable/26161568

Government of Alberta. (2020a). *Cut red tape*. https://www.alberta.ca/cut-red-tape.aspx

Government of Alberta. (2020b). *Wages and salaries in Alberta*. https://alis.alberta.ca/occinfo/wages-and-salaries-in-alberta/?offset=0&letter=all&sort=Title

Griffin, S. (1997). Ecofeminism and meaning. In K. Warren (Ed.), *Ecofeminism: Women, culture, nature* (pp. 213–226). Indiana University Pres.

Haider, S. (2016). The shooting in Orlando, terrorism or toxic masculinity (or both?). *Men and Masculinities*, *19*(5), 555–565. https://doi.org/10.1177/1097184X16664952

Haraway, D. (1988). Situated knowledges: The science question in feminism and the privilege of partial perspective. *Feminist Studies*, *14*(3), 575–599. https://doi.org/10.2307/3178066

Haraway, D. (1991). *Simians, cyborgs, and women: The reinvention of nature*. Free Association Books.

Haraway, D. (2016). *Staying with the trouble: Making kin in the Chthulucene*. Duke University Press.

Harding, S. (1991), *Whose science? Whose knowledge? Thinking from women's lives*. Cornell University Press.

Holter, Ø. G. (1997). *Gender, patriarchy and capitalism: A social forms analysis* [Doctoral thesis, University of Oslo]. Work Research Institute.

Hultman, M. (2013). The making of an environmental hero: A history of ecomodern masculinity, fuel cells and Arnold Schwarzenegger. *Environmental Humanities, 2*(1), 79–99.

Hultman, M., & Anshelm, J. (2017). Masculinities of global climate change: Exploring ecomodern, industrial and ecological masculinity. In M. Griffin Cohen (Ed.), *Climate change and gender in rich countries* (pp. 19–34). Routledge.

International Energy Agency. (2021). *Net zero by 2050: A roadmap for the global energy sector*. www.iea.org

Jenney, A., & Exner-Cortens, D. (2018). Toxic masculinity and mental health in young women: An analysis of 13 Reasons Why. *Journal of Women and Social Work, 33*(3), 410–417. https://doi.org/10.1177/0886109918762492

Johnson, O. W., Gerber, V., & Muhoza, C. (2019). Gender, culture and energy transitions in rural Africa. *Energy Resources and Social Sciences, 49*, 169–179. https://doi.org/10.1016/j.erss.2018.11.004

Jones, E., & Fionda, F. (2017, November 8). In search of Canada's elusive shadow population. *The Discourse*. https://thediscourse.ca/data/canadas-shadow-population.

Kaijser, A., & Kronsell, A. (2014). Climate change through the lens of intersectionality. *Environmental Politics, 23*(3), 417–433. https://doi.org/10.1080/09644016.2013.835203

Kenney, J. [@jkenney]. (2018, November 30). *Justin Trudeau says pipelines must go through a "gender based analysis" because male construction workers have "impacts." Darned right they [Tweet]*. Twitter. https://twitter.com/jkenney/status/1068582203225890816

Kimmel, M. (2017). *Angry white men: American masculinity at the end of an era*. Nation Books.

Krauss, C. (1993). Women and toxic waste protest: Race, class, and gender as resources of resistance. *Qualitative Sociology, 16*(3), 247–262. https://doi.org/10.1007/BF00990101

MacGregor, S. (2009). A stranger silence still: The need for feminist social research on climate change. *The Sociological Review, 57*(2), 124–140. https://doi.org/10.1111/j.1467-954X.2010.01889.x

McCright, A. M., & Dunlap, R. E. (2011). The politicization of climate change and polarization in the American public's view of global warming, 2001–2010. *The Sociological Quarterly, 52*(2), 155–194. https://doi.org/10.1111/j.1533-8525.2011.01198.x

Messerschmidt, J. W. (2018). *Hegemonic masculinity: Formulation, reformulation, and amplification*. Rowman & Littlefield.

Mies, M., Shiva, V., & Salleh, A. (2014). *Ecofeminism* (2nd ed.). Zed Books.

Miller, G. E. (2004). Frontier masculinity in the oil industry: The experience of women engineers. *Gender, Work & Organization, 11*(1), 47–73. https://doi.org/10.1111/j.1468-0432.2004.00220.x

Moskos, M. (2020). Why is the gender revolution uneven and stalled? Gender essentialism and men's movement into 'women's work'. *Gender, Work & Organization, 27*(4), 527–544. https://doi.org/10.1111/gwao.12406

National Inquiry into Missing and Murdered Indigenous Women. (2019). *Reclaiming power and place*. https://www.mmiwg-ffada.ca/final-report/

Natural Resources Canada. (2020). *Energy and the economy*. https://www.nrcan.gc.ca /science-data/data-analysis/energy-data-analysis/energy-facts/energy-and-economy /20062

Oparaocha, S., & Dutta, S. (2011). Gender and energy for sustainable development. *Current Opinion in Environmental Sustainability*, *3*(4), 265–271. https://doi.org/10 .1016/j.cosust.2011.07.003

Parent, M. C., Gobble, T. D., & Rochlen, A. (2018). Social media behavior, toxic masculinity, and depression. *Psychology of Men and Masculinity*, *20*(3), 277–287. https://doi.org/10.1037/men0000156

Peeples, J. A., & DeLuca, K. M. (2006). The truth of the matter: Motherhood, community and environmental justice. *Women's Studies in Communication*, *29*(1), 59–87. https:// doi.org/10.1080/07491409.2006.10757628

Plumwood, V. (2018). Ecofeminist analysis and the culture of ecological denial. In L. Stevens, P. Tait, & D. Varney (Eds.), *Feminist ecologies*. Palgrave Macmillan.

Pueyo, A., & Maestre, M. (2019). Linking energy access, gender and poverty: A review of the literature on productive uses of energy. *Energy Resources and Social Sciences*, *53*, 170–181. https://doi.org/10.1016/j.erss.2019.02.019

Romero, M. E. L. (2017, December 21). How boys and girls are taught different things about violence. *The Conversation*. https://theconversation.com/how-boys-and-girls-are -taught-different-things-about-violence-87779

Russell, E. L. (2021). *Alpha masculinity: Hegemony in language and discourse*. Springer Nature.

Ryan, S. E. (2014). Rethinking gender and identity in energy studies. *Energy Resources and Social Sciences*, *1*, 96–105. https://doi.org/10.1016/j.erss.2014.02.008

Ryder, S. S. (2018). Developing an intersectionally-informed, multi-sited, critical policy ethnography to examine power and procedural justice in multiscalar energy and climate change decision making processes. *Energy Resources and Social Sciences*, *45*, 266– 275. https://doi.org/10.1016/j.erss.2018.08.005

Schippers, M. (2007). Recovering the feminine other: Masculinity, femininity, and gender hegemony. *Theory and Society*, *36*(1), 85–102. https://www.jstor.org/stable /4501776

Scott, R. R. (2010). *Removing mountains: Extracting nature and identity in the Appalachian coalfields*. University of Minnesota Press.

Shriver, T. E., Adams, A., & Einwohner, R. (2013). Motherhood and opportunities for activism before and after the Czech Velvet Revolution. *Mobilization*, *18*(3), 267–288. https://doi.org/10.17813/maiq.18.3.t272388n60470456

Statistics Canada. (2014). *Victimization of Aboriginal people in Canada, 2014* [No. 85-002-X]. Statistics Canada Catalogue. https://www150.statcan.gc.ca/n1/pub/85-002 -x/2016001/article/14631-eng.htm

Statistics Canada. (2020a). *Employment by industry, annual*. https://www150.statcan.gc.ca /t1/tbl1/en/tv.action?pid=1410020201&pickMembers%5B0%5D=1.10&pickMembers %5B1%5D=2.1&cubeTimeFrame.startYear=2015&cubeTimeFrame.endYear=2019 &referencePeriods=20150101%2C20190101

Statistics Canada. (2020b). Table 11-10-0190-01 *Market income, government transfers, total income, income tax and after-tax income by economic family type*. https://doi.org /10.25318/1110019001-eng

Terry, G. (2009). No climate justice without gender justice: An overview of the issues. *Gender and Development, 17*(1), 5–18. https://www.jstor.org/stable/27809203

Thurton, D. (2017). *'Alarming number' of Suncor employees test positive for drugs and alcohol.* CBC. https://www.cbc.ca/news/canada/edmonton/suncor-drug-alcohol-test-1 .4424061

Van der Ploeg, F., & Rezai, A. (2020). Stranded assets in the transition to a carbon-free economy. *Annual Review of Resource Economics, 12*, 281–298.

Wilson, S. (2014). Gendering oil: Tracing western petrosexual relations. In R. Barrett & D. Worden (Eds.), *Oil culture* (pp. 244–263). University of Minnesota Press.

Women's Centre of Calgary. (2020). *Fast facts.* https://www.womenscentrecalgary.org/ work-for-change/fast-facts/

Wright, W. (2001). *The wild west: The mythical cowboy and social theory.* Sage.

# Reflection on Chapter 4

## Finding Balance: Gender, Extractive Industries, and Climate Change

*Mary Boyden*

*__Mary Boyden__ lives in the mining city of Timmins, Ontario, Canada. She is the Managing Director for Eighth Fire Solutions, a social enterprise established upon the direction of the AMAK Elders, a traditional council of Indigenous Elders. Led by the AMAK Elders, Mary facilitates Indigenous-led land reclamation strategies. Her work helps to bring the power of Traditional Ecological Knowledge to respond to critical contemporary and future challenges. For 26 years, Mary has collaborated with Keepers of the Circle, an organization led by Indigenous women who provide a range of programmes and services for Indigenous women in resource contexts. The organization is currently working to enhance meaningful participation of Indigenous women in impact assessment processes. Mary's passion for facilitating respect of Indigenous Ways of Knowing by mainstream society is fed by her own journey as the product of Indigenous and Settler ancestors who paved the way for Canada as we know it today. Mary has worked as an ally within Treaty 9 and beyond to build cross-sector relationships throughout her life.*

*In this conversational reflection, Mary shares her experiences as one of the first women to work underground in the Porcupine Camp. Drawing upon Anishinaabe traditional knowledge shared by her teachers, Mary discusses the contemporary situation of gender, extractivism, and climate change—and where the future may take us.*

**Amber:** *So, I'd like to first learn a bit about you. So, if you could tell me some of your background, your experiences, about yourself—and how you come to this topic of extractivism.*

**Mary:** I believe it's important to situate ourselves in any conversation we have, because people deserve to understand the basis for our perspectives. I grew up in Toronto, part of a big family. In the 1960s it was important to fit into society and to be socially engaged. Some aspects of our background were

DOI: 10.4324/9781003089209-9

never spoken out loud. Every year though, my dad would take us down into the northeastern United States, where his family originated, to visit a spiritual guide who was very influential for my parents. On these trips, we would go through the mountains, we would visit certain places, we would stop at cemeteries, and that was one of the ways that we got to know where we actually came from on his side of the family. And then on my mom's side of the family, we spent all our weekends, all our family time out on the land and water in Lake Simcoe, Georgian Bay, and the Trent Canal System. Those waterways taught us about our roots by being physically in those places.

Years later as a young mom and wife, I ended up living and working in Elliot Lake, which is a mining community, and I had the opportunity to go to college there. Uranium mining depends on ventilation to manage the radiation levels in the mines, and I was one of many women who worked in ventilation, engineering, and other trades there. Elliot Lake was very much a resource development town in those days, and everything centred on the mines. It was a very unique time and place because everything was possible. The mines were very large underground; everything was open, the drifts had lighting, and the main drifts were so large it was possible for haul trucks to pass each other.

My circumstances changed, and I ended up coming to Timmins to work in 1986, still working in ventilation in the underground environment. But gold mining is very different from uranium mining, both culturally and from a mining perspective. At that time in the 80s in Timmins, there were virtually no women underground; there were just none. I was actually the first woman underground in the mine that I worked at. I loved the environment and experienced so much more than just taking air samples. It was exciting, and I loved being underground, loved it. There is a spirit that's there, there is an energy that's there. I really enjoyed being in that energy. I lasted as long as I could, a couple of years, and then it just didn't work anymore because of social dynamics in the community. As a woman employed underground in those days, there was so much social stigma that it was even difficult to do things like open a bank account; I remember the woman who was supposed to be helping me open my bank account was so distraught that I was in the environment where her husband was working, that she said I was going to cause an accident.

I transitioned to the local college to work with women entering non-traditional trades. During those years, in the 90s, I started realizing that women and Indigenous people actually shared a lot of commonality, when it came to why they were being excluded from the best paying trades jobs. Thirty years later, today we see the same trends occurring related to the predominance of White males in the highest paying and most skilled occupations in mining. At that time, we tried to open doors for women by taking advantage of government programmes and partnering with feminist organizations even though our agendas were fundamentally different.

Fast forward again to 2008, just before Ontario's *Mining Act* was modernized to stimulate the need for mining companies and First Nations to have a relationship—a relationship that for the first time considered inclusion for Indigenous Peoples. I was given an opportunity to work for a local community, Wahgoshig First Nation, as they implemented their strategy to participate in the resource development taking place in their territory. After over 100 years of active mining in the region, this First Nation began to talk to the myriad of exploration, mining, and contractor companies who had been operating all around them. This work taught me so much about what a First Nation goes through, particularly a tiny and fairly remote community, as they negotiate in the interest of their members with international corporations. Timmins, which exists because of mining, has human and financial resources to position itself as the centre of resource development in the region, whereas the First Nations who have territory in and around the same developments have only been participating since 2010 and with very limited human and financial resources. So, all this time has gone by with no relationship. Communities like Wahgoshig have been totally on the outside looking in while also dealing with the violence that is part of opening up the land, interrupting the movement of people on the land, and putting women and children at particular risk.

By 2010, when the *Mining Act* changed, one of the pieces of the new puzzle was Impact Benefit Agreements. At this point, I started a job with a major mining company in the Timmins area, to bridge relationships between the company and the four rights-bearing First Nations in their area of operations. The company had very specific goals as to what they wanted. The communities knew what they wanted, and each one of them had a different way. We worked together for five years to build an agreement between the four communities and the company that was operating the mines, mines that had been in place for over 100 years at that point.

This was another very unique opportunity, where everything was possible because nobody had any previous experience in building relationships of this kind. A group of very specific Wisdom Keepers from across central Canada took a risk to come together to study what mining does to the land and agreed to work beside the company to help us all understand how the land and the people are connected in their healing. Engineers, geologists, miners, and environmentalists from the company, as well as the government and some universities, learned from the Wisdom Keepers and in turn began to change some of their practices to reflect the wisdom being shared. The focus for the work was mines that have not been closed properly; once mines have been operating for over 100 years in an active mining region, you get a lot of closure properties all over the place that just become part of the landscape. Environmental protections were not in place until the 1980s in Ontario related to the closure of mines.

**Amber:** So, the mines were left just as is, open?

**Mary:** Yes, and Timmins is built on these closed and abandoned mine properties. A closed mine means that there would be tailings. In mining, tailings are very similar to what they are in the oil sands. They're like a mixture of chemicals and ore left over from processing the ore. There is ground rock, but it's encased in all kinds of really toxic chemicals. So, they're just left in low-lying areas, or lakes, in liquid form. Eventually, after many years, they dry up and you start getting grass growing. We're built on that. We just take it for granted because of economics; we are a mining community.

With that comes these outcomes: It affects our water, affects the air, it affects our health as human beings. We're the end result of all that. Before us, it's the animals, it's the trees, it's the migrating birds, the geese who are landing on these beautiful big wet spots that are easy to land on. They're landing in the water that's sitting on top of these places, and then continuing their migration to the North where they are hunted.

We're also on the very edge of the watershed here, the Arctic watershed. All the water that we generate here is flowing to the James Bay coast, it's flowing north from here. All the impacts we're making are not just for us—they are much, much bigger. Mining is very much dependent on economics, right? It's all about economics, it's all about moving money from one part of the cycle to the next. Miners know how to start a mine, they know how to do the prospecting, they know how to do the geology, they know how to do the engineering. They are just beginning to learn how to look after the closure. They only really began that process in any real way here in Timmins in about 2004. There has been a hundred years of leaving these spaces behind. Then it became a big process to learn how to close them because by then the environmentalists were starting to see it, and social responsibility said it was not acceptable, and it was also part of the regulatory process. So they had to start. And now in Ontario, you can't open a mine without knowing how you're going to close it.

**Amber:** Okay.

**Mary:** In Timmins, it is part of our history. The burden of cleaning up historic mine sites is a liability to a company. On one hand, there is continued economy being stimulated by new mines and on the other hand there is a need to see the cumulative impacts of more than a century of mining on the environment and the people. And of course, all the other beings who share this land with us…the plant life, trees, waters, animals, birds, and fish. The Wisdom Keepers agreed to help us see ways to look at the land in a more balanced way by learning from the last part of the mining cycle. Many of us learned from these teachers and continue to see the world in a different way because of spending this time with them.

**Amber:** Can you tell me about gender relations in the mining industry today?

**Mary:** When I decided to move back into mining, I was working with a wise woman at that time. She was helping me understand myself—my connection to the world, and connection to myself. And she had shared with me a lot about the relationship between the feminine and the masculine and the historical connections of what had happened to us all…how the masculine became dominant and the feminine became the lesser, and how that became ingrained into all of us. She was the first one who helped me understand that we are all a combination of the masculine and the feminine energies. If we are a woman, then we carry more of that feminine energy. If we're men, we carry more of the masculine. But each one of us is in perfect balance, we carry it just like the earth, just like the whole universe.

Life is built from this balance of masculine and feminine. We need both. The violence of the last couple of thousand years literally evolved from the perception that the mind is the most powerful part of us, that when we are emotional, we are being weak. Men certainly can't cry. And women shouldn't cry either in a lot of situations because it's just not "appropriate." What she also helped me understand is that we have all become part of the same way of thinking, so it comes quickly in our judgements. It's a lot of work to undo those things. The ways we've been socialized to live, it takes a lot to unravel all that and to accept what balance actually means. Here on Turtle Island (North America) this way of living with the world changed with colonization and the need to take the resources from the land. Prior to being colonized, there was an inherent understanding of balance with all life.

The advice that she gave me was, when I go back into that environment, don't try to be a man. That environment is already unbalanced. The masculine influence is in everything from the economics to the engineering to the actual actions of mining. I learned that back in the 80s or 90s when we were working with women who were getting into trades. The first thing we did at that point was we taught them how to think like a man so that they would be able to survive. They needed to get desensitized when men are being violent or when certain comments are made…the dialogue that might happen in a lunchroom, for example. It's very cultural, but our job was to desensitize women to that so that they could get along.

It was the women who couldn't be desensitized who had a really hard time. They were the ones who had to stand up and say, "that's wrong, I don't agree with that. That's called sexual harassment." Because we wanted women to have all the opportunities in the world, it meant compromising some of what we really are or what we actually believe in.

And so my teacher said, "When you go back into that environment," she said,

> It needs you to bring the feminine energy into it… because it's out of balance, it's looking for balance. All of life is looking to be balanced. So it needs the feminine perspective, it needs the women in the environment in order to try to re-balance it. Don't go in acting like a man, thinking like a man, compromising yourself.

That was powerful. And it stuck with me. Especially because going into a management situation, it's all about strategizing, out-thinking the other, planning ahead, how are you going to weaponize the money you have, the power you have to get what you want.

Then those Wisdom Keepers, who I consider my teachers, also started bringing us the ceremonies. We had to begin to see the land as a living being. We are actually of the Earth, we are not superior to it. And that's the other thing that the male imbalance has brought, is the idea that we can dominate, we can extract, we can manage. What's going to happen in any situation, we can control it. And this is not what it's actually about—we need to work with it. There's always been extraction. Mining has always happened in different ways because the Earth is there to share. We are part of it, that relationship has always been here on Turtle Island. But, when we take anything to an extreme and pull it out of balance, then there are implications to that. And a lot of what these Wisdom Keepers did, over the six years that they worked with us, was to try to restore some of that balance.

On our watershed where, in a regulatory sense, we're only responsible for this mine, this operation, this is exactly what we're doing. Even though there's five other mines doing the same thing and the cumulative effects are so much greater because they're in a huge landscape, we are only responsible for our little piece of what I'm doing right now. No thinking about the future; very little consideration of all the past accumulation. It's very centric on the now. And it's very selfish, obviously. It's a very limited perspective.

And so the Elders started helping us understand some of these things in such a way that once you understand it, you can't un-understand it. Once you see it, you can close your eyes, and that might work for a while. But it can't last forever. Every one of us as human beings is autonomous, but we don't operate on our own, we need each other. Every one of us has a heart and a soul and a mind. But there comes a point when we're no longer just responsible for ourselves.

**Amber:** Could you tell me a little bit about how your work with Keepers of the Circle came about and maybe the work that the organization is doing in resource contexts?

**Mary:** My relationship with the Keepers of the Circle actually began 26 years ago, even though the project I am currently working with started 16 months ago. It's a small grassroots group of women from the local First Nations that formed in one little part of Ontario, the Temiskaming region. And they have grown to have impact across the country because of the vision of one woman and her sister, because of their connections within the community, their knowledge and experience and ability to learn from other ways of knowing. Keepers of the Circle stimulates growth for women. As soon as you stimulate the growth for a woman, you're influencing her children, you're influencing her partner, you're influencing her parents and everyone else. It's driven by that one opportunity for growth.

When you work with women, you can either just train them, or you can just provide social services to them, or you can just help them with childcare, or you can just look after their health needs. Keepers of the Circle has become a place where all those things are happening. It starts with women participating economically in a way that puts them onto the same level playing field as everybody else, but it also provides all the other aspects that we need in our lives, like supporting women who are going to go into a mining environment with the childcare that they need.

The work that we're doing on impact assessment evolved as an opportunity to take women beyond just participating economically, but to figuring out how we participate in the decisions that are going to be made about resource development.

There are a couple of really big influences that have informed our work at Keepers of the Circle related to impact assessment. One is the National Report on Missing and Murdered Indigenous Women and Girls. It's actually referenced in the chapter [Letourneau and Davidson]. There's a whole section in that report that directly relates the violence against the land to the violence against Indigenous women's bodies. The extreme rise in women going missing, women being violated, women being hurt in all kinds of ways…it's very real and it's very clear what's happening. One of the things that the report called attention to is the fact that it's very reactive right now; it's taken for granted that we're going to have the man-camp, we're going to have the contractors coming in, we're going to have all this happening and as a side effect, there is violence against the women and the children.

**Amber:** Yes.

**Mary:** The recommendation in the report is that we can plan for these disruptions instead of merely reacting to them. We don't need to wait until it happens. The services, the programmes…they can already be put into place at the grassroots before the resource is developed and before the contracts are awarded. And it's not like we're painting every contractor in the world with a brush, because that's not fair either. But the way we currently do resource development, there is a basic disregard of the sovereignty of every one of those beings that's being interrupted. Humans, the land, the water, the plants, the rocks themselves, the air that's surrounding all of it…it's each of those things. So when we start way further back, we understand that each one of these resource developments does not start at the process of impact assessment.

By the time an impact assessment occurs, there might have been two or three or four generations of people working towards making that project happen. All the work that happens before we get to the point where we're able to make that decision or when we're asked to participate in a decision is a huge economic driver. Indigenous women are not participating in the machine that ultimately dictates the opportunities that they may end up being given at the very end of a lengthy process that has impacted their way of life, safety, and wellbeing.

So, at Keepers of the Circle we're working on building tools for Indigenous women to participate more effectively in impact assessment processes throughout Canada. Part of that awareness means understanding the whole history, understanding how we've gotten to where we are now, understanding the power imbalance even within communities. The whole cycle ultimately begins with separating the original people from their land, and it is the women who embody the connection to the land. We need to face the reality of this and to begin the dialogues between us all.

There is a very strong teaching that really influences my way of seeing the world. It's an Anishinaabe prophecy, so it's based in central Canada here in North America. It is the teaching of the Seven Fires prophecy, and then the Eighth Fire, which is the time we are preparing for. We believe that we are now in that seventh time, we're in the Seventh Fire, which is the time when we need to make a decision as human beings about which road we are going to take to move forward. We are either going to choose a path of destruction, or we're going to choose that path of all people coming together in one great nation, to move forward as the human race, together. What is shared within the Seven Fires is that there are very clear responsibilities and gifts that were given to each of the four races of human beings. In the Medicine Wheel it's talked about as the White race, the Yellow race, the Red race, and the Black race.

That's all people on the earth, and each of those nations was put here on the Earth with specific gifts and with responsibilities through the cycle of history and time. Depending on our decision to choose life, we will need to prepare the fire that will sustain us as human beings on this planet. We can choose how to feed that fire. If we decide to feed that fire so that it kills us, that is our choice. If we decide to feed the fire so that it maintains what we need to live as human beings on this earth, that will also be our choice.

It also means understanding what each of us bring to this and giving space for the others to bring their knowledge in, too. I've heard wise people talk about the chaos that we're in right now: It has to do with that balance. We've totally forgotten the balance by staying on one path for too long that has taken us out of balance. We've also become very, very disrespectful of each other's knowledge, responsibilities, and gifts. It's painful to look at those things, to have those conversations, to unpack all that comes from being able to even think about working together.

We need the knowledge of all the people in the wheel to make the decisions, to find the new ways. We can do it. We will need to come together to figure those things out, and to have the conversations, to give each other the space and the respect and the acknowledgement and the gifting and the back-and-forth to create that possibility. Otherwise, we're choosing the other part of the path.

**Amber:** I think what you're saying also relates to climate change. The Seven Fires and the agency that we have, the choices we can make about the Eighth Fire and the future. And I think climate change fits into this because we are at a crisis point as well. Climate change is happening whether we want to think about it or not.

**Mary:** Exactly. Think of what we've been doing to violate the resources that are available to us as human beings to live here, without giving back. Mother Earth and all her resources are here for us to be part of, they're for us to use, they're for us to care for, they're for us to respect. That give-and-take must happen. With what's happening with climate change…it feels like we're on this bobsled racing down. And there's very little you can do, but there will be something that's going to stop us at the end. We can also choose at any point to roll off that bobsled. It's going to hurt for a bit, but I think we've got that chance. Our minds are capable of allowing us to see the possibilities, if we are led by our hearts. We're choosing all the time to dull ourselves down and not see it, not think about it.

It's also way more than that, though. It's bringing together the ways of understanding like we are doing with this interview. This particular process that we're in right now is academic-led. In the context of the Eighth Fire, "academic-led" means specifically that it's knowledge from the White part of the circle. And this part of the circle invites Red nation knowledge into this discussion. And what's the follow through to this? Our responsibility together is to make this conversation something that can be understood, that can be read, that can be used by people and not just for the sake of continuing to write things or put things down on paper. It's to share, it's to stimulate that dialogue and action.

You and I take the time to have this relationship because it is about more than any one of us. Climate change is not something external to ourselves. We have to understand that relationship and all the relationships that are resulting from it, right? With climate change, it's not just us who are being impacted. There are species that are disappearing every day right now. Whole species of beings who have a reason to be here on Earth and they're disappearing.

**Amber:** Yes.

**Mary:** Maybe it's about recreating our economy, the basis of our economy. It is the dollar. What if that basis was carbon sequestration? What if that was the basis of our economy? What if the human health and wellness index became the basis? Even if we don't all see all the other non-human beings on the earth as relevant and as equal to us, if not better, more than equal to us…if we can't see that? Okay, then what about our own human wellness? And using that as the basis?

**Amber:** You are speaking to what Angeline Letourneau and Debra Davidson are saying in the chapter as well. You've been talking about structures and how so much has happened before the moment of impact assessment—which has already set us up for a certain outcome. It's been so structurally embedded. In this chapter, the authors talk about how engrained masculinity and a certain economic structure are, in the oil industry culture. And what you're saying is so interesting because you're talking about how, by bringing in more ways of knowing, by bringing in balance, maybe we can start to shift toward other ways of knowing. And other ways of behaving in the natural world around us.

**Mary:** Exactly. That awareness happened in the work we did with the Wisdom Keepers. Individuals in the mining companies were influenced by that relationship with Elders, and they learned something different. Our geologists, there were a few of them...working with the Elders, learning about the culture, learning about that different way of seeing geology than what they had learned in university. It elevated something that was already in them, right? It allowed it to bloom. A lot of those people are doing other work in different places now. It just brought pieces together for themselves, seeing things that they didn't even realize we're really missing. And they've carried that forward.

I think the solution that we're coming to at the end of the day, at least part of the solution, is our understanding that this knowledge is already in different places. We can't know everything and we also don't need to categorize things in our way. Part of the communicating of the Eighth Fire is that knowledge is held in any part of that wheel or in any one of those nations or in any one of those corners of the world.

A big issue in resource development so far, and part of the damage that's happened for the people in the processes, the social processes we have, is that people's own knowledge is sometimes taken and used for other purposes, or categorized in a new way or studied by others, and in a way that it was never meant to be... And where the knowledge belongs, it's gone from there. Not just gone, it's taken. It's still held by its original keepers, but all of a sudden it's being used for other purposes out there. An example is when the medicine people are so protective of where certain medicines are growing, because they know that in the past, when something is shown to be a healing remedy, all of a sudden it's taken, marketed, and exploited.

So, the whole concept of resource development isn't just about the mind, isn't just about persons. It's about every aspect of how we end up living, or every aspect of our life. When things are merely extracted, including knowledge that is used disrespectfully, we've lost that give and take, we've lost the relationship, we've lost the respect. And again, within that Seventh Fire prophecy there was a point when that knowledge was taken and carefully put somewhere so that it couldn't be exploited. Now that we're trying to figure out how to have these conversations globally within these nations, towards all our healing, towards all our survival, that knowledge is ready to come out, but it will only come when it's going to be respected. And being respected means leaving it with the people that it belongs to and hearing it in the way that they are able to teach it.

# 5 Embodied Perceptions, Everydayness, and Simultaneity in Climate Governance by Spanish Women Pastoralists

*Federica Ravera, Elisa Oteros-Rozas, and María Fernández-Giménez*

**Federica Ravera** *is Ramón y Cajal Senior Researcher at the Department of Geography of the University of Girona (Spain). Her collaboration with communities and grassroots organizations has been influencing her transdisciplinary profile, which integrates approaches and methods from the natural and social sciences, and the arts, as well as from local and Indigenous knowledge systems to study the vulnerability and adaptation of rural agri-food systems to global change. Specifically, Federica's line of research focuses on the analysis of socio-institutional innovations, collective actions, and the role of traditional and local knowledge in responding to global changes in (silvo)pastoral systems of the Mediterranean context and high mountain regions (Andes, Himalayas, and Pyrenees). Through a perspective of feminist political ecology, her research is mainly oriented to understand the dynamics of power that create inequalities, and the differential conditions that create new opportunities in the transformations of social-ecological rural systems.*

**Elisa Oteros-Rozas** *is Ecologist by background and a postdoctoral researcher at the Chair on Agroecology and Food Systems of the University of Vic (Spain) and the FRACTAL Collective of feminist women researchers in social-ecological systems. She works, both as an academic and as an activist, in Agroecology and Food Sovereignty, with a particular focus on pastoral and transhumance systems. She has paid attention to traditional and local ecological knowledge around livestock farming and, lately, with a particular focus on gender, women, and new peasantries in pastoralism. Other recent work deals with pastoralists' perceptions of and adaptations to climate change and coexistence with wildlife.*

DOI: 10.4324/9781003089209-10

*María E. Fernández-Giménez leads the Rangeland Social-Ecological Systems (RSES) lab in the Department of Forest & Rangeland Stewardship at Colorado State University. She works closely with ranching and pastoralist communities in the western US, Spain, and Mongolia to support sustainable and equitable rangeland management through knowledge co-production and research. She studies how ranchers/pastoralists make decisions, individually and collectively, and how their decisions affect rangeland social-ecological systems. Recent work focuses on traditional ecological knowledge, community-based rangeland management, and gender and social justice in rangeland and pastoral systems.*

## Introduction

### *Climate Change, Livestock Management, and Gender*

The relationship between climate change and livestock farming is twofold (Rojas-Downing et al., 2017). On the one hand, according to the Intergovernmental Panel on Climate Change (IPCC), livestock, particularly industrialized and intensive production systems, are deepening climate change by fostering land use change and deforestation, and by emitting 14.5% of greenhouse gasses (IPCC, 2019). On the other hand, climate change negatively affects extensive livestock farming and pastoralism by reducing primary productivity (i.e., forage resources), decreasing water availability, increasing the frequency of extreme weather events (particularly in arid and semi-arid regions), and spreading disease, among other Processes (e.g., Muchuru & Nhamo, 2017; Rubio & Roig, 2017). In this chapter, we use both the terms "pastoralism" and "extensive livestock production" to refer to the livestock management systems where animals meet most of their nutritional needs by grazing natural pastures and permanent grasslands or silvopastoral systems outdoors, usually in small or medium scale family-run enterprises[1]. Nomadic or transhumant pastoralism (i.e., long- and short-term mobility of herds) is also adopted by some farmers and communities to respond to the uncertainty, scarcity, or variability of access to resources (fresh forage and water) during certain seasons.

The projections of climate change in Spain suggest that Mediterranean ecosystems are close to a tipping point, requiring urgent responses (Guiot & Cramer, 2016). Increasing temperatures, decreasing precipitation, and increasing extreme events (e.g., heatwaves, prolonged drought, changes in seasonality, and rising wildfire risks, among others) are expected to lead to progressive aridification of certain areas and a "Mediterraneanization" of others, directly affecting the distribution of plant species, health of animals, and local livelihoods. Spanish extensive livestock management and pastoralism, which include a wide range of landscapes from the threatened *dehesas* (anthropogenic oak savannahs), to agroforestry systems and mountain grasslands, cover more land area and encompass more diverse management systems than any others in Europe (Herrera, 2020). According to the Spanish National Climate Change Adaptation Plan (Rubio & Roig, 2017), traditional extensive livestock management practices are powerful adaptive responses to a variable climate and may help conserve ecosystems in the face of new climate change hazards. Additionally, the fifth IPCC report demonstrates the potential for carbon storage in pasturelands (IPCC, 2015). However, Spain's traditional pastoralism is also threatened by climate change, and multiple environmental hazards and concomitant drivers amplify the risk of an irreversible crisis. Indeed, to understand the complex effects of climate change in the Spanish context, they must be considered in relation to other multiple stressors including depopulation and land abandonment, land use change, and agricultural policy and cultural changes (Camarero & Sampedro, 2019).

Although gender dimensions of climate change in pastoralist systems are recently receiving increasing attention, such studies have mainly focused on the Global South (e.g., Aregu et al., 2016; Balehey et al., 2018; Omolo, 2010; Venkatasubramanian & Ramnarain, 2018). Studies of the Global North remain scarce (e.g., Buchanan et al., 2016; Reed et al., 2014; Wilmer & Fernández-Giménez, 2016a, b) or absent, as in the case of Spanish extensive livestock systems. Given the potentially critical role that women play in the future of Spain's extensive livestock management systems, our research aims to address this gap through a case study of women pastoralists' experiences of, and responses to, climate change in Spain.

### *Feminist Scholarship and Climate Change*

In this chapter we draw on three main bodies of feminist literature applied to climate change studies: Embodied scholarship, the analysis of "everyday" experiences and resistance (following the concept introduced by Smith in 1987), and the concept of interlocking systems of oppression (i.e., simultaneity or intersectionality) as recently applied by critical feminist political ecologists.

Embodiment scholarship increasingly challenges the disembodied and masculinist science behind climate change discourse and policymaking at broad scales and illuminates the implications of climate change in local places. Feminist political ecology (FPE) literature points out that little research addresses the socio-political dimensions of climate change (MacGregor, 2010), such as how perceptions, impacts, and responses differ among subjects. FPE further highlights how climate change biophysical studies often marginalize the voices of the most vulnerable people (e.g., Indigenous people, women) (Buechler & Hanson, 2015). Additionally, FPE scholars argue that contemporary Western societies often view "climate change" as an abstract phenomenon, despite framing it in language of urgency and emergency. The global scale of climate change contributes to this abstraction, but so does the prevalence of scientific and technical discourse and a knowledge system, based on data and biophysical information on climate change, that downplay more experiential, embodied, and non-scientific forms of perceiving and feeling it. Post-humanist and new-materialist scholars investigate trans-corporeal climate change that places the problem, and thereby its solutions, within and on our bodies: "Rain might extend into our arthritic joints, sun might literally color our skin, and the chill of the wind might echo through the hidden hallways of our eardrums" (Neimanis & Walker, 2014, p. 560). Following these authors, we consider climate change as an embodied "subject" that interacts with our bodies and emotions. Specifically, we apply the concept of "weathering"; that is, "a particular way of understanding how bodies, places and the weather are all inter-implicated in our climate-changing world" (Neimanis & Hamilton, 2018, p. 80).

A second body of literature also informs our study. Following the concept of "everyday life and relations" (Smith, 1987), we employ a feminist lens to explore climate governance, i.e., the range of strategies, actions, and measures of climate change mitigation and adaptation. This literature emphasizes the importance of more closely considering women's standpoints and their everyday spaces, and their practices and relations around climate change that produce and regulate subjects and subjectivities, affecting people's daily lives. Examples of this literature are the work of Bee (2013, 2014) and Bee et al. (2015), who illustrate how climate change becomes something not only visceral, material, and embodied, as we mentioned above, but also part of everyday lives. Examining women's everyday spaces and mundane experiences is a way to understand how gendered power relations shape women's ways of mitigating and adapting (or not) to climate change and related hazards.

We thus adopt a situated ethnography, which includes semi-structured life history interviews, participant observation, and workshops, and a review of social media messages to capture both 1) how shepherdesses and women livestock operators physically and emotionally feel climate change; and 2) how women pastoralists in their daily lives respond and adapt to climate and environmental change. This qualitative research facilitates collecting situated knowledge and stories from voices of subjects who are not usually heard in the first person—i.e., shepherdesses and women livestock operators—as an alternative to the "official" voices—i.e., scientists and men pastoralists. These women make a different sense of the world, as sources of knowledge for understanding and responding to changes in rural societies. Nevertheless, we do not view women as unitary subjects. As suggested by Neimanis and Hamilton (2018), the phenomenon of weathering is situated and not all bodies react similarly. Experiences of climate change are embedded in social and political contexts. Thus, weathering is a concept based on the politics of differences and intersectionality.

A third body of literature from new FPE and intersectionality informs our study. With specific reference to the Global North, an emerging academic literature examines how gender, race, class, education, and place, among other categories of social difference, make for differential perceptions, experiences and responses of rural people to climate change and environmental disasters more generally (e.g., drought, flooding, fire) (e.g., Alston, 2006; Buechler & Hanson, 2015; David & Enarson, 2012; Fletcher, 2018; García Gonzalez, 2019; Rodó-de-Zarate & Baylina, 2018; Walker et al., 2021; Whyte, 2014). Developed through the work of Black feminists, such as the Combahee River Collective and Crenshaw (1991), the concept of intersectionality has been described as the overlapping and interactive ways that gender and race can oppress (or empower) specific individuals or groups (Crenshaw, 1991). The concept calls for "simultaneity" or "interlocking" conjunctures of common histories and oppressions and disjunctures that produce differentiation between subjects, knowledges, experiences (Johnson, 2017).

Aware of the different use of intersectionality and/or simultaneity from its origins and applications in Black feminism, we follow new FPE scholars in using these concepts to analyze the intertwined categories of social difference that influence women's abilities to use gendered environmental knowledge and gain political voice and decision-making power at various levels of social organization (Elmhirst, 2011; Nightingale, 2011). In this way we also contribute to empirical research on intersectionality and/or simultaneity in climate change studies in the Global North (Ravera et al., 2016; Reed et al., 2014). In the Spanish context, and specifically studying pastoralist systems, Fernández-Giménez et al. (2021) recognize that women pastoralists express common histories of oppression, but also represent a diversity of experiences, motivations, challenges, and learning processes that vary greatly based on a woman's age, origins, education, social class, pathways into livestock husbandry, locality, and role in the family and livestock enterprise.

Additionally, the analysis of intertwined oppressions is concerned not only with the local, but with the structures of power that permeate all social relations from the individual to the global (Winker & Degele, 2011). Different policies that address climate change and other multiple stressors with impacts on environmental hazards have both negative and positive effects on different groups. Studying normative assumptions beyond institutional arrangements and policies can help us to understand intersections of power. It is, thus, pertinent in our work to ask how structural forms of inequity are reproduced, reinforced, or challenged by socio-cultural norms, ideologies, and dominant discourses that determine which social locations possess the knowledge and value (or not) in shaping environmental, and, specifically climate, governance (Kaijser & Kronsell, 2014).

In this chapter, we apply these three key concepts from the literature—embodiment, everyday life, and simultaneity/intersectionality—to study how Spanish women engaged in extensive livestock production perceive and feel climate change; how their subjective, diverse, lived, and sometimes contradictory experiences are linked to adaptation and mitigation responses to climate change and in general to systemic changes; and how intersecting axes of power, from the individual to the institutional, shape climate change governance.

Moreover, by engaging participants in making meaning of their experiences and their feelings in the face of climate change, we seek to destabilize knowledge hierarchies among researchers and social actors. We adopt an ethnographic approach, collecting life histories of shepherdesses and women livestock operators and organizing discussions with pastoralist women via participatory workshops. As feminist researchers, our situated ethnography emphasizes transparency and reflexivity regarding our own positionalities and seeks to foster collaborative and reciprocal relationships between researchers and research participants (Baylina Ferré, 2004; England, 1994).

## Women in Extensive Livestock Management in Spain: A Situated Ethnography

### The Spanish Context

Extensive livestock production in Spain is an ancient and still-relevant practice that shapes rural landscapes and provides multiple ecosystem services and cultural benefits (e.g., Azcarate et al., 2013; Montserrat Recoder, 2009; Oteros-Rozas et al., 2014). In 2015, livestock farming in Spain emitted 86.4 million tons of $CO_2$ equivalent, more than half of which is attributable to the agricultural production of animal feed and is therefore associated primarily with industrial animal production systems (Herrera, 2020). However, recent research highlights the caveats of the accounting methodologies used by the IPCC, which do not include a baseline scenario with wild herbivores also emitting greenhouse gasses (Manzano & White, 2019), or the carbon sink that grasslands and other natural pastures contribute (Freibauer et al., 2004). The impacts of climate change on Spanish grasslands have been extensively reported and include the loss of plant diversity, reduced soil moisture availability (the "Mediterraneanization" of the northern Iberian Peninsula and the aridification of the Southern area), rapid changes in species distribution that are driving extinctions and a resulting biotic homogenization, edaphic changes, changing wildfire regimes, increased frequency and impacts of extreme weather events like drought, or changes in seasonality that affect plant phenology and pollination, among others (Rubio & Roig, 2017). These ecological changes are already having positive and negative effects on extensive livestock management in Spain, with negative effects including reduction of animal production, new and increased diseases, reduced animal wellbeing, the loss of livestock breeds, and positive effects including reduction in the neonatal mortality of lambs, goats, and calves (Medina, 2016; Rubio & Roig, 2017).

In Spain, women have long been part of livestock management systems, although their roles and visibility varied by region (García-Ramón, 1989). Since the mid-20th century, livestock production systems in Spain largely shifted away from traditional extensive management and towards intensive industrial livestock farming (Guzman et al., 2018), leading to rural depopulation, masculinization, and land abandonment across large areas (Camarero & Sampedro, 2008, 2019; González Díaz et al., 2019). Today, however, Spanish women appear to be increasing their role in the livestock sector as both salaried and non-salaried family labour, nearly reaching parity with the number of men in the sector (FADEMUR, 2011).

### Methods

Fieldwork and co-interpretation of findings for this study took place from June 2018 to August 2020. During this time, the three co-authors collected life-histories of a diverse range of women directly and indirectly involved in extensive livestock management, including women who own or co-own operations or who work with livestock as family members or employees of an operation owned by someone else. We focused the interviews in four geographic areas: Andalucía (southern Spain), the Northwest (Zamora, León, Asturies, and Cantabria), the central Pyrenees and lowlands of Aragó, and Catalunya. We identified interviewees (n = 31) through existing research and personal contacts, a country-wide network of Spanish women pastoralists, Ganaderas en Red (GeR), and a regional network, Ramaderes de Catalunya (Ramaderes.cat). Interviewees ranged from 22 to 96 years old. Following a situated ethnographic method, we visited most participants' operations or accompanied participants during their daily work. Several interviews involved extended participant observation or repeated interactions. Participants shared about their origins and families; how they entered the livestock sector; the nature of their farm, livestock, and land management; their motivations and goals; and the difficulties and barriers they faced. They also reflected on past and future changes in the sector and the environment, the main drivers and challenges they perceive, and on present and future responses. To enrich our material and capture collective discourse, we also analyzed the Facebook and Twitter accounts of the two associations we worked with: @ganaderasenred and @ramaderescat. We reviewed the last two years (2019 and 2020) of Tweets and messages, searching for references to key words such as climate emergency, climate change, fires, drought, mitigation, adaptation, disaster, Gloria (i.e., Hurricane Gloria), etc.

We coded the interviews and Twitter posts looking for references to drivers of change and challenges, and to adaptation, mitigation, and transformation at different scales. We also coded feelings and emotions we found in the interviews related to drivers and impacts of changes. Finally, we coded references to social categories and power dynamics.

We convened in-person workshops with interview participants and other women pastoralists in Andalucía (n = 11 participants), Northwest Spain (n = 11), Catalunya (n = 5) and the Aragón Pyrenees (n = 3) and Catalan Pyrenees (n = 16) to discuss preliminary research findings with an expanded group of participants, collect additional data on women's experiences and perspectives, and engage participants in data interpretation. We organized an additional virtual meeting with participants in August 2020 to share and further discuss our evolving analysis. Workshop participants discussed how to use the findings to advance their goals, such as increasing empowerment and visibility of pastoralist women in their families, communities, and the sector, improving rural services, and educating society about extensive livestock production. This research was conducted with the free, prior, and informed consent of the participants under Colorado State University's Institutional Research Ethics Board [350-18H]. Names of interview and workshop participants are pseudonyms. We did not anonymize Twitter posts as they are already publicly accessible.

## Findings and Discussion

### *Embodied Perceptions of Climate Change and Environmental Hazards*

When "wildfires of sixth generation" arose in the Iberian Peninsula in 2019 and 2020, "the body of firemen fought against them as well as the bodies of shepherds and shepherdesses felt them" (GeR, 2020a).

There is the fire. Come quickly—your animals are burning. If there is a taboo in the rural, it's the fire. Nobody names it. Since we arrived with our project of the Las Cumbres [farm operation] five years ago, we have thought to establish somewhere a protocol, but we didn't do it, to not invoke the devil. On July the 31st of 2020 the devil appeared. […] and the fear, lot of fear. [and then] The smell. The silence. To walk the land, your land, when it burned, it's unforgettable. Where have the birds gone? This year arrived to remind us we are so fragile. (Lucía, Huelva)

Shocking experiences like Lucía's (which are recalled by herself in the next chapter of the present volume) are repeatedly recounted by women in social media and our conversations. An older woman pastoralist in a video recorded in July 2020 also told about her feelings of fear, sadness, and helplessness facing fires—a natural phenomenon of Mediterranean ecosystems, which has intensified in Spain in the last decades. Such intensification (i.e., fires are less frequent but much more devastating, unpredictable, and larger) is explained by the interaction of aggravating causes, including climate change and resulting water stressed forests, the expansion of unmanaged forests related to rural depopulation, and resulting abandonment of traditional farming activities like grazing and controlled burns that limit shrub expansion. Similar to those two stories, during one of the workshops with shepherdesses in Catalunya, women expressed their worries of being the most vulnerable to wildfires and their fear due to lack of knowledge and tools to face them.

Wildfires are just one of the environmental hazards linked to climate change that women pastoralists portray through their everyday feelings. In January 2020, the Ramaderes de Catalunya Twitter feed shared images of snow, ice, and water disasters provoked by Hurricane Gloria, which interfered with the daily life of women and animals—another example of women's lived experiences with multiple and uncertain climate change effects. Our interviews also include many references to prolonged droughts, floods, landslides, and other intense climate-related events. The following Twitter post epitomizes the theme of embodiment—of experiencing, feeling, and knowing change through emotion ("our heart") and bodies ("our skin"):

The dehydration of soils is chronic. The springs, the mountain and the planet know, our skin and our heart, of men and women farmers who live and work in the rural, know. Who produces food see and feel what means the lack of water. It's impossible for shepherdesses to not see our mountains and pastures losing springs at an irrevocable speed. (GeR, 2020b)

These stories remind us of feminist scholars' argument for the importance of "pluralistic politics of knowledge for effective climate governance" (Bee et al., 2015, p. 9). Women farmers express different ways of knowing climate hazards. Far from suffering a "hyper-hypo-affective disorder" (Colebrook, 2011, p. 45), like much of Western society, they feel their own bodies engaged in the rhythm of everyday life with a weather-world in flux, experiencing emotions in the face of wildfires, droughts, extreme weather events, and natural disasters. As Neimanis and Walker (2014, p. 573) suggest: "These records, memories, and intensities are indications of 'insurgent vulnerabilities': we are responsive to the weather, as it is to us." Women express, then, their ethos of responsiveness to such urgency, because they are "weathering" their bodies and time.

In one example of this ethos of responsiveness, "Ganaderas en Red" in September 2019, joined the climate strike, because, in their own words: "We are part of the planet, who is our ally and we are aware of the havoc of climate change in pastures and springs" (GeR, 2019a). A few months later at the COP25, where they participated in a roundtable on extensive livestock management and pastoralism, they also communicated their concerns about the effects of climate change on ecosystems and biodiversity and, ultimately, on their animals, their livelihoods, and physical and mental health (GeR, 2019b).

Despite their consciousness and concerns about climate change, other emotions are also expressed by women pastoralists, like rage to have to combat the image—promoted by international reports on climate change (e.g., IPCC, 2019) and social media—that livestock production as a major polluter and contributor to greenhouse gases, without distinguishing between intensive industrial livestock production and small-scale place-based extensive livestock management, and to showcase the environmental and health benefits of the latter. In December 2018, a long post on @ramaderescat manifested their disapproval of an article titled "If you want to save the planet, don't eat meat" (ARA, 2018). In response, they wrote:

> Our ovaries are swollen and we have decided to react.... It's not that we don't agree with the fact that we should eat less meat... but can we put a little emphasis on what meat and vegetables we need to eat if we want to 'fight climate change'? (Thus, we say) if you want to save the planet, choose the right system that produces the food you eat. (Ramaderes de Catalunya, 2018)

This tweet, with a passionate reaction to the newspaper article, also highlights how the climate emergency is intertwined with a particular agri-food model grounded in a set of agrarian, land use, and urbanization policies that further drive both intensification and rural depopulation and land abandonment, ultimately harming ecosystems and biodiversity. As suggested by several scholars that focus on the politics of climate change, it is difficult to determine which local impacts are, and which are not, directly attributable to climate change rather than to indirect drivers of climate change and underlying stressors that contribute to rural vulnerability (Räsänen et al., 2016). Several women during the interviews highlighted their consciousness of how rural depopulation and the abandonment of livestock farming, especially sheep and goat herds, contribute to shrub encroachment and increase fire risk and biodiversity loss. Some of them are also critical about how policies oriented to tourism overlook rural population and extensive livestock management's needs, and therefore are inadequate to prevent rural abandonment and pasture loss. They expressed their sadness, worries, and fears on the future of villages without people and pastoralism: "I would not like for the world to see the house closed, the village empty. That is something I cannot bear. That is something that gives me a lot of pain... and my tears fell" (Rosa, North of Spain).

Moreover, some women are claiming the new EU Common Agricultural Policy (CAP) should consider who is working the land and mitigating climate change by fixing $CO_2$, maintaining habitat diversity, fertilizing soils, and renewing nutrient cycles. "The peasantry, we are among the main landscape managers. What resources will be devoted to the advice and support to adapt the agroecosystems to climate change? We want a landscape capable of guaranteeing Food Sovereignty, now and in 10 years" (Ramaderes de Catalunya, 2019).

However, some of them also express anger about how the EU CAP is not currently oriented to pastoralism's conservation and promotion:

> We have everything certified organic, which requires even more pastures, not for them to produce and survive, mind you! We want access to the subsidies of the CAP, the damn CAP, [but it's] the poisoned candy... we need even more pastures than animals need to survive. And we don't have any economic and social recognition. We are adding value for settlement of the people, for the generational turnover, for the self-esteem, for the environment ... all these (services) should be economically taken into account in the new CAP. (Sandra, Andalusia)

### *Exploring Everyday Spaces and Lived Experiences of Adaptation to, and Mitigation of, Climate Change*

Following Smith's concept of "everyday" (1987), the life histories of women tell us about the mundane, ordinary activities and practices in homes, neighbourhoods, and communities through which women respond to climate change and other concomitant drivers and stressors.

Our findings show that women pastoralists' routines, experiences, knowledge, and spaces are never unimportant, because they shed light on how women express their perceptions and capacity to respond to climate change, and how they resist, conserving their landscapes and ways of life while they also transform the system.

Our interviews revealed a variety of everyday and mundane practices women undertake that sustain rural communities and cultural landscapes, from one woman's dedication to restoring and maintaining the physical infrastructure of her village, to many participants' commitment to keeping locally adapted rare and heritage livestock breeds (Fernández-Giménez et al., 2021). Though knowledge and adoption of practices for conservation of landscapes and biodiversity are not exclusive to women, many interviewees defended the use of small, controlled burns and grazing in mountains to reduce shrub encroachment and wildfire risk, and to manage horizontal and vertical connectivity. In Spain, the warming and drying climate combines with land use change, especially abandonment of formerly grazed or cultivated lands, to increase wildfire risk. The progressive abandonment of traditional agricultural and livestock management activities of the Spanish rural territories since the 1970s has resulted in increased forest and shrub cover—e.g., around 690,000 hectares from 1994 to 2006 according to Benayas et al. (2007), and therefore increased wildfire hazards.

Traditionally, shepherds in some regions used small controlled (or "artisanal") burns to keep pastures cleared of shrubs and accessible for grazing (Fernández-Giménez & Fillat Estaque, 2012). Flocks of goats and sheep that graze on the new growth of woody plants also helped manage the highly flammable shrubs and understory vegetation and reduced risk of large intense fires. Several participants, like Carla and Juana below, referenced the role that their livestock play in maintaining the landscape and reducing the wildfire risk.

> Here it's a strange balance, because the climate is so changeable. There are drought years when you don't have grass, or tiny amounts of grass, but there are years when you have a lot of grass and you need to have sheep to eat it because if you don't there is a huge fire risk. You must have sheep. (Carla, Andalucía)

> The sheep do it, but the goats much more. Do you know what a goat can clear? It's outrageous. If you go there to a field, the goat is always on the hillside, eating the trees, the shrubs… It helps the mountains a lot. (Juana, Pyrenees)

Today, in many locations, the government prohibits controlled burns, and livestock numbers (especially sheep and goats that favour woody species) are insufficient to keep shrubs at bay. Since the economic crisis of the late 2000s, there has been a decline in public investment in expensive mechanized forest thinning and management. Our participants frequently commented on the important roles of livestock and controlled burning in mitigating shrub encroachment and wildfires. Some, like Marina in the following excerpt, expressed frustration at an administration that promotes a distorted image of pastoralists and their role in wildfire management:

> We burn because it is a management tool that we have used all our lives, and the people from the cities when they see the fire say 'delinquents, the stockmen are burning again,' and no, we are cleaning the mountains so the houses don't burn, because they are going to burn any day because of the government's disastrous management. (Marina, Northwest)

Nine of the interviewees use the traditional practice of transhumance, i.e., repeated movements between distinct seasonal pasture areas in different geographical and ecological zones. As such, they help to maintain a practice that provides multiple environmental benefits (Oteros-Rozas et al., 2014), and uses mobility to adapt to variable environmental conditions over space and time (Oteros-Rozas et al., 2013). Victoria, a transhumant from northwestern Spain discusses the value of the practice for having open spaces and controlling fires as well as for seed dispersal, as expressed below. Although she does not link seed dispersal to climate change adaptation, the role of sheep in transporting propagules could serve as an inadvertent form of "assisted migration," a strategy to conserve biodiversity in a changing climate (Manzano & Malo, 2006).

> So there are plants that would be lost if the livestock didn't carry them, plants that are going to disappear. ...There are millions of seeds they say the animals expel along the [transhumant] route, as they graze from one side to another. In the village where we go in the spring one woman said that new plants are arriving that they have never seen before. (Victoria, Northwest)

Although women have long supported transhumance as members of transhumant families, until recently, it was relatively rare for women to take a visible and active role in the everyday activities of animal husbandry during the transhumance trips. One of our eldest participants described how she was rebuked by other local women when, in the 1970s, she first dressed in farmers' work clothes and went out to help her husband with the traditional "men's work" with the sheep. Her lifelong persistence in flouting some conventional gender roles while conforming to others exemplifies acts of embodied everyday resistance to dominant gendered relations of power in transhumant operations. Younger transhumant women like Juana also reported similar experiences:

> The first year I went up with transhumance, the whole trip [...], I was 13 years old, I remember that everyone was telling me: 'where are you going, you disgrace, you will not arrive, where are you going, where are you going, no way you can arrive...' [laugh] And I arrived, perfectly.

Juana is currently one of the youngest transhumant women in Spain.

Several women also mentioned specific practices related to soil management that aimed to increase carbon capture or reduce greenhouse gas emissions. For example, Sandra from Andalucía discussed the importance of good grazing management to limit erosion and maintain critical soil functions like infiltration in the context of climate change. This excerpt also highlights the detailed ecological knowledge that women expressed in discussing their daily management activities:

> Every time now with climate change, it rains more rarely but torrentially. If you have the soil, if you have it not bare but protected, it will act like a sponge and prevent the loss of the fertile topsoil, and furthermore, on top of that it soaks [the rain] up, holds it and filters it. (Sandra, Andalucía)

Most of the previous examples highlight women's roles in maintaining or advocating for traditional practices that also function as responses to climate change. Such examples also reveal women pastoralists' nuanced ecological knowledge. As such, women's everyday activities and knowledge are often naturalized and under-recognized, while affording them daily and direct experiences with resource scarcity, natural disasters, and environmental changes. We shift focus towards women's roles as innovators and proactive agents of change, linking their actions to a climate change consciousness. In our study, a number of women implement novel approaches to climate change adaptation and mitigation, as well as actions to transform extensive livestock production more broadly towards greater social and environmental sustainability. Many (but not all) of these women are young and/or newcomers to the livestock sector. Those who are most politically engaged with the food sovereignty movement propose a new agro-social paradigm shift, that is, a shift to a new food system strongly connected to the environmental and climate change movements. These findings align with previous studies on newcomers in the European Union (Monllor i Rico & Fuller, 2016; Pinto-Correia et al., 2015).

Marina from the Northwest exemplifies an innovative approach to climate change mitigation. Marina won an international prize for women entrepreneurs for a project that aimed to reduce methane emissions from cattle waste, while limiting nitrate pollution of soils and aquifers, increasing soil fertility, and improving livelihoods. In the interview, she first described the challenges of manure management and her concern about releasing contaminants into the aquifer.

Other interviewees reported working on projects at the enterprise, regional, or sectoral levels that directly or indirectly mitigate climate change through technology adoption or institutional innovations, often while also working towards a transformation of agri-food systems and the livestock economy. For example, at the enterprise level, several women installed solar panels on their farms. In interviews, they linked this choice to their growing awareness of the interconnections between their lived experiences of climate change and the greenhouse gas emissions of their agricultural activities.

Several interviewees also reported innovations in value-added local branding and marketing of their animal products (e.g., locally produced lamb, charcuterie, and cheese) and linked them to climate change mitigation. Erika, a young shepherdess from Catalunya, connected these efforts to the need for a "zero kilometer" agri-food system that directly decreases greenhouse gas emissions. In sparsely populated rural Spain, livestock slaughtering and packing facilities are often located far from the small villages where herders live. Erika is working for more local food systems, collaborating with others to advocate for the introduction of mobile slaughterhouses that better fit both farmers' and consumers' needs and concerns regarding animal wellbeing and the reduction of transportation emissions.

Several women also recounted innovative examples of public-private collaborations in Catalunya where livestock are used to decrease wildfire risk. For example, in the project Ramats de Foc, shepherds access land and graze the forests, providing mutual benefits for young pastoralists, who often face difficulties in securing access to grazing lands, and the local government administration, for whom cost-effective fire risk reduction is a priority. The Artisan Butchers Guild of Girona and participating restaurants add value to the products of participating pastoralists, through a specific label: "Turning our forests into silvopastoral systems. Not only does this remove fuel from the forest floor [it also] turn[s] it into tasteful and nutritious meat" (Ramats de Foc, 2020).

Finally, the increasing participation of women in sectoral organizations and the rise in women pastoralists' networks, facilitated by social media and digital communication, indicate women's growing roles in transforming the extensive livestock sector. To our knowledge, the first such virtual network was Ganaderas en Red (GeR), founded in 2016 by a group of women pastoralists and women advocates for pastoralism, who were tired of attending meetings about the extensive livestock sector where women's perspectives and voices were ignored or not represented. Several other groups have spun off from the initial network, including THE one specific to Catalunya (Ramaderas.cat).

In addition to increasing women's visibility and including their voices at the sector level, GeR has served as a mutual support and knowledge exchange network for women pastoralists across Spain. A clear example of this was the initiatives set in place during the COVID-19 pandemic, in which the network fostered internal consumption and helped advertise direct sales. All the GeR participants interviewed mentioned how these virtual and in-person spaces support women in processing knowledge and information, including on environmental change, within their households and communities, and in sharing it beyond their localities. Our findings thus align with others from feminist sociology and geography that suggest that women in rural contexts specifically pursue women's cooperation and networks as new strategies to respond to the masculinization of rural settings and knowledge systems (e.g., Ní Fhlatharta & Farrell, 2017; Porto Castro et al., 2015; Sachs et al., 2016).

## *Simultaneity in Spanish Pastoralism in the Face of Urgent Responses to Climate Hazards and Other Multiple Stressors*

As Kaijser and Kronsell (2014) remark, differentiated situated knowledge and understanding related to climate hazards and responses is derived from women's position in society, their background, and the place in which they live. An intersectional approach seeks to avoid oversimplifying women's experiences by attributing their knowledge and experiences to a universal aspect of being female. Previous studies (Fernández-Giménez et al., 2019) show how Spanish women pastoralists still experience barriers to accessing resources. In this work, we recognize simultaneous and interacting social positions related to women's different origins, education, and pathways into livestock management that influence differential knowledges, attitudes, and discourses around natural resources management and drivers of change.

Women interviewed in this study differ in age, family status, and origin (rural versus urban), which affects their access to land and knowledge of livestock management. Although each woman's life history is unique, we identified three main pathways through which women enter the livestock sector (Fernández-Giménez et al., 2021). Some women inherited the profession from their families, others married into the business, and some were newcomers who reported starting "from zero." Three of them are hired shepherdesses who work for other livestock operators. Nearly two thirds of the interviewees were married or partnered in heterosexual relationships, while one third were unmarried, widowed, or divorced. Two-thirds were mothers, including three with young children.

In our study, intersectionality emerges from 1) issues of stereotyping and discrimination of those who did not come from a pastoralist background, especially young women; 2) tensions between shepherdesses versus urban environmental and animalist activists.

## The Intersection of Origin, Age, and Gender

A general discourse recognizes as pivotal the repopulation of rural spaces and revitalization of the extensive livestock sector in the face of climate hazards. Newcomers have capabilities related to their often-urban origins and education—most of them come from agrarian and forestry schools and academic training—but they lack access to material resources, especially land and animals, and to inherited knowledge and lived experience with animal husbandry. Newcomers often come to the countryside seeking a more tranquil and grounded quality of life and/ or a context for child-rearing compared to their urban origins. They also more frequently articulate specific ideological motivations related to agroecology, food sovereignty, and ecofeminism in their discourses and practices of environmental governance. Paradoxically, in their own words, they are advantaged by not carrying with them a long history of difficulties in rural villages and the livestock sector.

In contrast, despite their inherited access to land and animals (in many cases), and their rich traditional knowledge related to livestock management, women from rural agricultural family backgrounds more frequently express a negative discourse on how to deal with the multiple stressors of the livestock sector. Several expressed fatigue from defending the rural world, and specifically a traditional extensive livestock sector "on its way to extinction" (Fernández-Giménez et al., 2022).

In some cases, our interviewees are pioneering solutions to overcome such unequal distribution of resources, as in the case of Cloe, who encountered barriers to entering the livestock sector as an immigrant and newcomer starting from zero after arriving from Latin America. Her project represents a novel arrangement involving access to land within a protected area to graze forested areas for fire control. The project was sponsored by a land bank that supports young people returning to rural communities and farming livelihoods. She also created a mobile farm and cheese factory, which she calls a "circus farm," that moves according to availability of grazing land.

Our findings also illustrate how social relations are shaped by power and how the access to, and control over, decision making is not only gender-differentiated, but also depends on intersecting identities (Bee et al., 2015). With some exceptions, Spanish women pastoralists still face barriers to holding leadership positions and implementing innovative practices and policies within formal livestock organizations. Such barriers are related to age, origin, and social status (like being a hired shepherdess rather than livestock owner), in addition to gender. As observed by other authors (Fernández-Giménez et al., 2022; Oteros-Rozas et al.,

2017), established pastoralists sometimes ostracize, discourage, or express skepticism about newcomers, especially if they are young women. Laia's experience is one example, where she described how, initially, other shepherds did not include her in decisions: "[At the beginning] for old people of this mountain, who don't know you because you come not from here, it's so strange a young girl from abroad who works as employee in a farm as a shepherd." (Laia, Catalunya)

Several newcomers recounted how they fought such discrimination within livestock organizations and expressed their resistance by negotiating for the recognition of their voices. These results are similar to those of Baylina Ferré et al. (2015) who studied newcomers from urban areas who resettled in rural Galicia and Catalunya. Although some of the landless (and animal-less) young women who worked as hired shepherds described their precariousness, they also expressed their sense of freedom. Also, they are very active in advocating for improved working conditions.

*Gendered Aspects of the Urban-Rural Power Divide*

To address intersectionality in light of climate change, it is also important to analyze how social and political institutions may "take part in the construction and reinforcement of injustices and intersectional categorizations" Kaijser and Kronsell (2014, p. 426). In the context of Spanish pastoralism, these dynamics are most visible in the ongoing tensions between rural agrarian society and urban areas that are home to large populations and centres of power and governance.

First, conservation policies have excluded from decisions those who are suffering the main impacts of environmental hazards in the Spanish countryside, and who hold possible adaptation and mitigation responses to these hazards. Such policies are designed, under certain representations of the problem of environmental change, by a hegemonic national/urban centre in contrast to a rural periphery. An example is found in the conflicts that have arisen in the last decades between extensive livestock management and other competing land uses, such as tourism and biodiversity conservation. Embedded, but hidden, in these conflicts is the divergence in understandings of human-nature relationships and of rural and urban values, worldviews, and power relations. Beginning in the 1980s, conservation policies and strategies on protected areas and key species (re)introduction have been designed largely from the urban centres' desktops, supported by conservation organizations with their headquarters mostly in metropolitan areas (Beltran & Vaccaro, 2014; Vaccaro & Beltran, 2009). Such policies have most negatively affected small farmers, but little has been studied about potential gender differences in attitudes towards such policies or relationships to wildlife. In Spain, women typically do not engage in legal or illegal predator control. In the northwestern workshop women verbalized emotions evoked by the conflict between livestock farming and growing wolf populations, mentioning fear, helplessness, rage, and disappointment. They committed to invite the government administration to come and experience the everyday reality of landscape and livestock management. They further committed to a social media campaign to raise awareness about the need to consider the needs and benefits of extensive livestock management, not only of wildlife. Overall, historically, local communities, and specifically women, have not been included in decision-making processes and progressively have been dispossessed of their legitimate governance over their own territories and management of local resources. Here, Marina captures the feeling of such conflict around the conservation of wild fauna by the government and environmental organizations:

> [Names of two environmental organizations] have not even sat down to negotiate. Nor to talk, when we were talking about the management of wolves. And they have not sat down to dialogue. So, with those positions, not even sitting down, there is already a confrontation. I organized a meeting and more than 250 livestock farmers attended, a unique opportunity to see both parts… and those didn't even want to sit down. So those people [are] radical, but they have power and they have very powerful lobbies, and the only thing that they are doing is killing us. So, my question is: who has more rights, wildlife or people? (Marina, Northwest)

Women interviewed, especially those from traditional rural backgrounds, described how the rural world has been neglected and undervalued by urban culture. As the following quotes suggest, intertwined power relations related to culture and origins within the urban/rural continuum and centre-periphery dynamics, normatively reproduce in public opinions the narrative that defines which experiences have value and who has the legitimate technical and scientific knowledge to take decisions. Specifically, urban-centred "environmentally and climate-friendly" movements are set in contrast to a "negative other," represented by extensive livestock managers who oppose conservation policies as they are currently designed and applied.

> What I dislike the most is sometimes the incomprehension of people who don't understand that you are here because you like it. That sentence of 'oh, you who have studied [at university], and you're here caring for cows.' And it's, I am taking care of cows because I want to care for cows. And the incomprehension that they don't realize that without a stockman or a farmer, they won't eat. They don't value you. I feel very empowered as a pastoralist, but when people don't see that the primary sector is fundamental and they go to you like, 'you there with the cows, washing off the dung,' and they diminish you. (Sara, Northwest)

The complexity rises as climate policies have direct and indirect effects on food policies and on food habits. The special report on climate change and land by the IPCC in 2019 described plant-based diets as a major opportunity for mitigating and adapting to climate change—and included a policy recommendation to reduce meat consumption. These recommendations, together with the increasingly popular climate emergency youth movement and the anti-speciesism movement, have also permeated the feminist movement, fuelling a conflict between urban, highly educated and academically supported anti-speciesism feminists and livestock farmers—specifically women pastoralists—some of whom also self-identify as feminists. In this sense, this urban/rural tension is gendered, because women shepherdesses have been treated differently in this public discourse of anti-speciesism than men.

On 8 March 2019, representatives of the urban feminist movement from Catalunya launched a manifesto in which anti-speciesism was among the values raised as central to feminism. In response, Ramaderes de Catalunya issued a public declaration expressing how they, as rural feminists, did not feel represented by such a manifesto, denouncing the supremacist attitude of urban feminists, explaining their ecofeminist perspective, and making visible the differences between intensive and extensive livestock production systems. This resulted in yet another newspaper article published by a number of highly educated women, from universities and anti-speciesism collectives, titled "Feminism must be anti-speciesism." Since these exchanges, ethical and environmental arguments are being discussed in various fora, creating an extremely vivid public debate laden with emotion from both sides. Meanwhile, both the feminist and the environmental movements in Spain seem to be increasingly incorporating veganism and the critique of livestock farming among their pillars. Anti-speciesism and vegan movements accuse women pastoralists of incoherence for their enslavement and discrimination against animals and especially non-human females. To reinforce this narrative and influence public opinion, anti-speciesism movements describe livestock management as one of the main causes of climate change, without any distinction between intensive industrial and globalized management and extensive management.

Women pastoralists are defending their right to self-identify as feminists, even when, and precisely because, their livelihood system depends on animal management. The engagement of some of the women interviewed within feminist environmental and food sovereignty movements led them to expand the discussion on animal wellbeing and work together with environmental and academic organizations on the linkage of livestock management, climate change and, more recently, zoonotic epidemic[2]. Around this topic, in contrast to conservation, the extensive livestock sector and part of the environmentalist movement are working as allies to publicly differentiate between production systems, supported by scientific evidence (del Prado & Manzano, 2020; Ecologistas en Acción, 2019; Herrera, 2020). Specifically, these alliances work to point out the inaccuracies of the current emissions accounting systems and to publicly defend a sustainable food system that takes care of people, animals, and the environment. Several of the interviewed women express frustration with these increasingly dominant discourses, as well as assumptions about environmental cause-and-effect relations at the root of current social-ecological challenges.

> But the problem is that look, within 20 years, what will this be? If we are going to be in the hands of the animal rights activists and the conservationists …And the future scares me because we will fall in the hands of those children

we are educating. I already told you about my son's biology book that says 'extensive livestock production pollutes.' No, no, it's the macro-farms that pollute, and emit gas. But my sheep that go through the mountains eating and pooping, this is far from pollution. It is scientifically certain that the extensive livestock production produces emissions, but it likely has lots of benefits. Someone has to explain this. (Sara, Northwest)

A final remark in this sense is that the analysis of implicit and explicit assumptions underlying these constructions tells us how the relationships between humans and nature are lived and defined, as well as how and to whom the responsibility for environmental protection is assigned and how it is related to other axes of power, such as being women, young, newcomers but living and working in marginalized rural spaces and sectors. This message is well illustrated by Cloe, a young migrant newcomer:

It seems foolish to me, to come here and set fire to my farm [metaphori-cally], while wearing Decathlon shoes made by exploited Thai girls, living in a place where you get your food from the supermarket and you don't question where it comes from or how it contributes to climate change. I sometimes feel that maybe this way of life [i.e., livestock farming] makes us less anthropocentric, that is, we are one more thing, we are equal to the sheep, to the birds, to everything. And we are one gear more. (Cloe, Catalunya)

## Conclusions

In the Spanish extensive livestock sector, women assume important roles in the management of resources, animals, and territories, despite their invisibility and undervaluation. Climate change affects the ecology of natural pastures and influences the livestock sector and livelihoods of Spanish rural society. Women pastoralists express their understanding and experience of climate hazards through their bodies, spaces (physical, virtual, symbolic), and emotions.

Avoiding a simplistic focus on vulnerability, interviewed women express their differential agency in adapting to and mitigating climate hazards in a variety of ways. On one hand, their everyday actions and mundane experiences express their sense of responsibility to conserve rural landscapes and village spaces, and to continue the traditional management practices that maintain biodiversity, soil, and pasture quality, prevent wildfires and other environmental hazards, and feed society with healthy food. As observed by Dowsley (2007) in her studies on Inuit women, our interviewees perform key roles in managing traditional knowledge, discussing and processing information about environmental change within the household and the community, and sharing it through social media.

On the other hand, many of the interviewees reject stereotypical images of rural womanhood. Specifically, they challenge the predefined roles of women as "helpers" in rural families, communities, and society, and they act as change agents, advancing multiple climate change adaptation and mitigation innovations at household, community, sectoral, and societal scales. Through a bricolage of elements that they have at hand, several of them deliberately create opportunities for tangible and intangible changes to address the problem. For instance, they implement new technologies, knowledge, land and animal management practices; forge novel private-public alliances against wildfires and towards ecological objectives; strengthen women's networks for collective support and knowledge exchange to deal with environmental changes; and work to shift social paradigms and values about the rural world by promoting the role of extensive pastoralism in healthy and environmental sustainable food production.

However, the politics of difference and intersectionality influence climate governance. In Spain, multiple barriers still obstruct transformative paths of adaptation to climate change, and these barriers are linked to intersecting social positions, each subject having distinct advantages and disadvantages. Specifically, as suggested by Fernández-Giménez et al. (2022) newcomers with non-rural and non-livestock backgrounds and limited access to land and animals, young women, and single women often experience the greatest barriers to entry in the extensive livestock sector.

Finally, we argue that drawing attention to the simultaneity of causes of marginality and power inequities among individuals and groups depending on their social location, and on the assumptions that privilege certain experiences and knowledges over others, may open a window of opportunity for transforming social relations and collaborating to build more inclusive solutions to climate hazards. Our situated ethnography supported relationship- building, mutual learning, and knowledge bridging between us, as researchers, and women pastoralists, shifting the focus on climate change from the biophysical to a political arena. Moreover, the research moved from the application of an analytical strategy of intersectionality towards simultaneity as critical praxis of inclusiveness that may help transformations. Additionally, the women's networks, GeR and Ramaderes de Catalunya, are examples of arenas that facilitate relationship-building, mutual support and care, and knowledge-sharing among women of different backgrounds and political ideologies. As such they work towards resolving conflicts, making visible the junction of power dynamics and interlocking institutional oppressions in order to fight against them, increasing women pastoralists' participation and voice in decision making, and engaging with common objectives. As suggested by feminist scholars, creating alliances based on common interests and solidarity rather than on fixed identities implies a move beyond merely identifying power patterns in a certain context to looking for common ground for action and engagement (Kaijser & Kronsell, 2014; Lykke, 2010; Mohanty, 2013).

Despite this initial analysis, we recognize that intersectionality is an understudied topic in our study context, and much work remains to explore the experiences of LGBTQ individuals, migrants, and other intersecting social identities in Spanish pastoralism.

## Notes

1 Anne Horsin, Claire Lebras, Jean-Pierre Theau. 2019. Extensive livestock production: Definition. Dictionnaire d'Agroecologie, https://dicoagroecologie.fr/en/encyclopedia/extensive-livestock-production/
2 See a recent article written by @ramaderescat "In defense of the pastoralism": https://directa.cat/en-defensa-del-pastoreig/.

## References

Alston, M. (2006). "I'd like to just walk out of here": Australian women's experience of drought. *Sociologia Ruralis*, *46*(2), 154–170. https://doi.org/10.1111/j.1467-9523.2006.00409.x

ARA. (2018, December) *Si vols salvar el planeta, menja menys carn.* https://www.ara.cat/premium/vols-salvar-planeta-menja-menys_1_2705580.html

Aregu, L., Darnhofer, I., Tegegne, A., Hoekstra, D., & Wurzinger, M. (2016). The impact of gender-blindness on social-ecological resilience: The case of a communal pasture in the highlands of Ethiopia. *Ambio*, *45*, 287–296. https://doi.org/10.1007/s13280-016-0846-x

Azcarate, F. M., Robleno, I., Seoane, J., Manzano, P., & Peco, B. (2013). Drove roads as local biodiversity reservoirs: Effects on landscape pattern and plant communities in a Mediterranean region. *Applied Vegetation Science*, *16*, 480–490.

Balehey, S., Tesfay, G., & Balehegn, M. (2018). Traditional gender inequalities limit pastoral women's opportunities for adaptation to climate change: Evidence from the Afar pastoralists of Ethiopia. *Pastoralism*, *8*, 1–14.

Baylina Ferré, M. B. (2004). Metodología para el estudio de las mujeres y la sociedad rural. *Estudious Geográficos*, *254*, 5–28. https://doi.org/10.3989/egeogr.2004.i254.190

Baylina Ferré, M. B., García-Ramón, M. D., Porto Castro, A. M., Rodó-de-Zárate, M., Salamaña Serra, I., Villarino Pérez, M. (2015). Género, trabajo y sostenibilidad de la vida en el medio rural. In J. de la Riva, P. Ibarra, R. Montorio, & M. Rodrigues (Eds.), *Análisis espacial y representación geográfica: innovación y aplicación* (pp. 1929–1936). Universidad de Zaragoza-AGE.

Bee, B. A. (2013). Who reaps what is sown? A feminist inquiry into climate change adaptation in two Mexican ejidos. *ACME: An International Journal for Critical Geographies*, *12*(1), 131–154. https://acme-journal.org/index.php/acme/article/view/955

Bee, B. A. (2014). "Si no comemos tortilla, no vivimos:" Women, climate change, and food security in central Mexico. *Agriculture and Human Values*, *31*, 607–620. https://doi.org/10.1007/s10460-014-9503-9

Bee, B. A., Rice, J., & Trauger, A. (2015). A feminist approach to climate change governance: Everyday and intimate politics. *Geography Compass*, *9*(6), 339–350. https://doi.org/10.1111/gec3.12218

Beltran, O., & Vaccaro, I. (2014). Between communal herding and state parcellation: The conflicting territorialities of the Spanish Pyrenees. In A. C. Dawson, L. Zanotti, & I. Vaccro (Eds.), *Negotiating territoriality: Spatial dialogues between state and tradition.* https://doi.org/10.4324/9781315813103. Routledge.

Benayas, J. M. R., Martins, A., Nicolau, J. M., & Schulz, J. J. (2007). Abandonment of agricultural land: An overview of drivers and consequences. *CAB Reviews: Perspectives in Agriculture Veterinary Science, Nutrition and Natural Resources*, *2*, 1–14. https://doi.org/10.1079/PAVSNNR20072057

Buchanan, A., Reed, M. G., & Lidestav, G. (2016). What's counted as a reindeer herder? Gender and the adaptive capacity of Sami reindeer herding communities in Sweden. *Ambio, 45*, 352–362. https://doi.org/10.1007/s13280-016-0834-1

Buechler, S., & Hanson, A. M. (Eds.). (2015). *A political ecology of women, water and global environmental change.* Routledge.

Camarero, L., & Sampedro, R. (2008). Why are women leaving? The mobility continuum as an explanation of rural masculinization process. *Revista Española de Investigaciones Sociológicas, 124*, 73–105. https://doi.org/10.2307/40184907

Camarero, L., & Sampedro, R. (2019). Despoblación y ruralidad transnacional: Crisis y arraigo rural en Castilla y León. *Economía Agraria y Recursos Naturales, Spanish Association of Agricultural Economists, 19*(1), 59–82. http://dx.doi.org/10.7201/earn .2019.01.04

Colebrook, C. (2011). Earth felt the wound: The affective divide. *Journal for Politics, Gender and Culture, 8*(1), 45–58. https://doi.org/10.51151/identities.v8i1.251

Crenshaw, K. (1991). Mapping the margins: Intersectionality, identity politics, and violence against women of color. *Stanford Law Review, 43*(6), 1241–1299. https://doi .org/10.2307/1229039

David, E., & Enarson, E. (Eds.). (2012). *The women of Katrina: How gender, race and class matter in an American disaster.* Vanderbilt University Press.

de Cumbres de Enmedio-Huelva, L. [@ganaderasenred]. (2020, August 20). *There is the fire. Come quickly: Your animals are burning.-If there is a taboo in the rural, it's the [Tweet].* Twitter.

del Prado, A., & Manzano, P. (2020). *La ganadería y su contribución al cambio climático.* Amigos de la Tierra & BC3 Basque Centre for Climate Change. https://www.tierra .org/wp-content/uploads/2020/09/Informe-Ganaderia-Cambio-climatico-Amigos-de-la -Tierra.pdf

Dowsley, M. (2007). Inuit perspectives on polar bears (Ursus maritimus) and climate change in Baffin Bay, Nunavut, Canada. *Research and Practice in Social Sciences, 2*(2), 53–57.

Ecologistas en Acción. (2019). *Agroecología para enfriar el planeta.* Ecologistas en Acción Creative Commons. https://www.ecologistasenaccion.org/wp-content/uploads /2019/12/informe-agroecologia-2019.pdf

Elmhirst, R. (2011). Introducing new feminist political ecologies. *Geoforum, 42*(2), 129–132. https://doi.org/10.1016/j.geoforum.2011.01.006

England, K. V. L. (1994). Getting personal: Reflexivity, positionality and feminist research. *Professional Geographer, 46*(1), 80–89. https://doi.org/10.1111/j.0033-0124 .1994.00080.x

FADEMUR. (2011). Situación actual de las mujeres rurales en España. *La Tierra Cuadernos, 17*, 28–34.

Fernández-Giménez, M. E., & Fillat Estaque, F. F. (2012). Pyrenean pastoralists' ecological knowledge: Documentation and application to natural resource management and adaptation. *Human Ecology, 40*, 287–300. https://doi.org/10.1007/s10745-012 -9463-x

Fernández-Giménez, M. E., Oteros-Rozas, E., & Ravera, F. (2019). *Co-creating knowledge for action with women pastoralists in Spain.* Asociación Trashumancia y Naturaleza.

Fernández-Giménez, M. E., Oteros-Rozas, E., & Ravera, F. (2021). Spanish women pastoralists' pathways into livestock management: Motivations, challenges and learning. *Journal of Rural Studies, 87*, 1–11. https://doi.org/10.1016/j.jrurstud.2021.08.019

Fernández-Giménez, M. E., Ravera F., & Oteros-Rozas, E. (2022). The invisible thread: Women as tradition-keepers and change-agents in Spanish pastoral social-ecological systems. *Ecology and Society 27*(2):4. https://doi.org/10.5751/ES-12794-270204.

Fletcher, A. J. (2018). More than women and men: A framework for gender and intersectionality research on environmental crisis and conflict. In C. Fröhlich, G. Gioli, R. Cremades, & H. Myrttinen (Eds.), *Water security across the gender divide* (pp. 35–58). https://doi.org/10.1007/978-3-319-64046-4_3. Springer.

Freibauer, A., Rounsevell, M. D. A., Smith, P., & Verhagen, J. (2004). Carbon sequestration in the agricultural soils of Europe. *Geoderma*, *122*(1), 1–23. https://doi.org/10.1016/j.geoderma.2004.01.021

Ganaderas en Red [@ganaderasenred]. (2019a, September 27). *We are part of the planet, who is our ally and we are aware of the havoc of climate change [Tweet]*. Twitter.

Ganaderas en Red [@ganaderasenred]. (2019b, December 7). *[Tweet]*. Twitter.

Ganaderas en Red [@ganaderasenred]. (2020a, July). *The body of firemen fought against them as well as the bodies of shepherds and shepherdesses felt them [Retweet]*. Twitter.

Ganaderas en Red [@ganaderasenred]. (2020b, October 13). *The dehydration of soils is chronic. The springs, the mountain and the planet know, our skin and our heart, of [Tweet]*. Twitter.

García González, M. (2019). Interseccionalitat i estudis de gènere en geografia rural: un estat de la qüestió (2008–2015). *Doc. d'Anàlisi Geogràfica, 65*(3), 603–627. https://doi.org/10.5565/rev/dag.575

García-Ramón, M. D. (1989). Actividad agraria y género en España: una aproximación a partir del Censo Agrario de 1982. *Documents d'Anàlisi Geogràfica, 14*, 89–114. https://doi.org/10.1234/no.disponible.a.RACO.41458

González Díaz, J. A., Celaya, R., Fernández García, F., Osoro, K., & Rosa García, R. (2019). Dynamics of rural landscapes in marginal areas of northern Spain: Past, present, and future. *Land Degradation & Development, 30*(2), 141–150. https://doi.org/10.1002/ldr.3201

Guiot, J., & Cramer, W. (2016). Climate change: The 2015 Paris Agreement thresholds and Mediterranean basin ecosystems. *Science*, American Association for the Advancement of Science, *354*(6311), 465–468. https://doi.org/10.1126/science.aah5015

Guzman, G. I., de Molina, M. G., Fernández, D. S., Infante-Amate, J., & Aguilera, E. (2018). Spanish agriculture from 1900 to 2008: A long-term perspective on agroecosystem energy from an agroecological approach. *Regional Environmental Change, 18*, 995–1008. https://doi.org/10.1007/s10113-017-1136-2

Herrera, P. M. (Ed.). (2020). *Livestock farming and climate change: An in-depth approach.* Fundación Entretantos and Plataforma por la Ganadería Extensiva y el Pastoralismo.

Intergovernmental Panel on Climate Change. (2015). *Climate change 2014: Synthesis report. Contribution of working groups I, II, and III to the Fifth assessment report of the intergovernmental panel on climate change* (R. K. Pachauri & L. A. Meyer, Eds.).

Intergovernmental Panel on Climate Change. (2019). *Climate change and land: An IPCC special report on climate change, desertification, land degradation, sustainable land management, food security, and greenhouse gas fluxes in terrestrial ecosystems* (P.R. Shukla, J. Skea, E. Calvo Buendia, V. Masson-Delmotte, H.-O. Pörtner, D. C. Roberts, P. Zhai, R. Slade, S. Connors, R. van Diemen, M. Ferrat, E. Haughey, S. Luz, S. Neogi, M. Pathak, J. Petzold, J. Portugal Pereira, P. Vyas, E. Huntley, K. Kissick, M. Belkacemi, & J. Malley, Eds.). World Meteorological Organization and United Nations Environment Programme.

Johnson, A. W. (2017). Resituating the crossroads: Theoretical innovations in Black feminist ethnography. *Souls*, *19*(4), 401–415. https://doi.org/10.1080/10999949.2018.1434350

Kaijser, A., & Kronsell, A. (2014). Climate change through the lens of intersectionality. *Environmental Politics*, *23*(3), 417–433. https://doi.org/10.1080/09644016.2013 .835203

Lykke, N. (2010). *Feminist studies: A guide to intersectional theory, methodology and writing*. Routledge Advances in Feminist Studies and Intersectionality.

MacGregor, S. (2010). Gender and climate change: From impacts to discourses. *Journal of the Indian Ocean Region*, *6*(2), 223–238. https://doi.org/10.1080/19480881.2010 .536669

Manzano, P., & Malo, J. E. (2006). Extreme long-distance seed dispersal via sheep. *Frontiers in Ecology and the Environment*, *4*(5), 244–248. https://doi.org/10.1890/1540 -9295(2006)004[0244:ELSDVS]2.0.CO;2

Manzano, P., & White, S. R. (2019). Intensifying pastoralism may not reduce greenhouse gas emissions: Wildlife-dominated landscape scenarios as a baseline in life-cycle analysis. *Climate Research*, *77*(2), 91–97. https://doi.org/10.3354/cr01555

Medina, F. (2016). *Impactos, vulnerabilidad y adaptación al cambio climático en el sector agrario: Aproximación al conocimiento y prácticas de gestión en España*. Oficina Española de Cambio Climático. Ministerio de Agricultura, Alimentación y Medio Ambiente.

Mohanty, C. T. (2013). *Feminism without borders, decolonizing theory, practicing solidarity*. Duke University Press.

Monllor i Rico, N., & Fuller, A. M. (2016). Newcomers to farming: Towards a new rurality in Europe. *Documents d'Anàlisi Geogràfica*, *62*(3), 531–551.

Montserrat Recoder, P. (2009). *La cultural que hace el paisaje*. Sociedad Española de Agricultura Ecologica.

Muchuru, S., & Nhamo, G. (2017). Climate change and the African livestock sector: Emerging adaptation measures from UNFCCC national communications, *International Journal of Climate Change Strategies and Management*, *9*(2), 241–260. https://doi.org /10.1108/ IJCCSM-07-2016-0093

Neimanis, A., & Hamilton J. M. (2018). Weathering. *Feminist Review*, *118*(1), 80–84. https://doi.org/10.1057/s41305-018-0097-8

Neimanis, A., & Walker, R. L. (2014). Weathering: Climate change and the "thick time" of transcorporeality. *Hypatia*, *29*(3), 558–575. https://doi.org/10.1111/hypa.12064

Ní Fhlatharta, A. M., & Farrell, M. (2017). Unravelling the strands of 'patriarchy' in rural innovation: A study of female innovators and their contribution to rural Connemara. *Journal of Rural Studies*, *54*, 15–27. https://doi.org/10.1016/j.jrurstud.2017.05.002

Nightingale, A. J. (2011). Bounding difference: Intersectionality and the material production of gender, caste, class and environment in Nepal. *Geoforum*, *42*(2), 153–162. https://doi .org/10.1016/j.geoforum.2010.03.004

Omolo, N. (2010). Gender and climate change-induced conflict in pastoral communities. Case study of Turkana in north-western Kenya. *African Journal of Conflict Resolution*, *10*(2), 81–102. https://www.accord.org.za/ajcr-issues/gender-and-climate-change -induced-conflict-in-pastoral-communities/

Oteros-Rozas, E., Ontillera-Sánchez, R., Sanosa, P., Gómez-Baggethun, E., & Reyes-García, V., González, J. A. (2013). Traditional ecological knowledge among transhumant pastoralists in Mediterranean Spain. *Ecology and Society*, *18*(3), Article 33. https://doi.org/10.5751/ES-05597-180333

Oteros-Rozas, E., Martin-Lopez, B., Gonzalez, J. A., Plieninger, T., Lopez, C. A., & Montes, C. (2014). Socio-cultural valuation of ecosystem services in a transhumance social-ecological network. *Regional Environmental Change, 14*, 1269–1289. https://doi .org/10.1007/s10113-013-0571-y

Oteros-Rozas, E., Rude, K., & Diaz Reviriego, I. (2017). ¿Who learns from whom? Social networks of knowledge exchange between new and local pastoralists (No. 85). In Elikadura21 (Ed.), *The future of food and challenges for agriculture in the 21st century: Debates about who, how and with what social, economic and ecological implications we will feed the world.* http://www.reverdea.com/wp-content/uploads/85-Oteros-Rozas.pdf

Pinto-Correia, T., Gonzalez, C., Sutherland, L. A., & Peneva, M. (2015). Lifestyle farming: Countryside consumption and transition towards new farming models. In L. A. Sutherland, I. Darnhofen, L. Zagata, & G. Wilson (Eds.), *Transition pathways towards sustainability in European agriculture* (pp. 67–82). CABI.

Porto Castro, A. M., Villarino Pérez, M., Baylina Ferré, M., García Ramón, M. D., & Salamaña Serra, I. (2015). Formación de las mujeres, empoderamiento e innovación rural. *Boletín De La Asociación De Geógrafos Españoles, 68*, 385–406. https://doi.org /10.21138/bage.1867

Ramaderes de Catalunya [@ramaderescat]. (2018, December 20). *Our ovaries are swollen and we have decided to react.... It's not that we don't agree with the fact that [Post].* Facebook.

Ramaderes de Catalunya [@ramaderescat]. (2019, September 18). *The peasantry, we are among the main landscape managers. What resources will be devoted to the advice and support to [Tweet].* Twitter.

Ramats de Foc [@ramatsdefoc]. (2020, March 24). *Turning our forests into silvopastoral systems. Not only does this remove fuel from the forest floor and turn it into [Tweet].* Twitter.

Räsänen, A., Juhola, S., Nygren, A., Käkönen, M., Kallio, M., Monge Monge, A., & Kanninen, M. (2016). Climate change, multiple stressors and human vulnerability: A systematic review. *Regional Environmental Change, 16*, 2291–2302. https://doi.org/10 .1007/s10113-016-0974-7

Ravera, F., Iniesta-Arandia, I., Martín-López, B., Pascual, U., & Bose, P. (2016). Gender perspectives in resilience, vulnerability and adaptation to global environmental change. *Ambio, 45*, 235–247. https://doi.org/10.1007/s13280-016-0842-1

Reed, M. G., Scott, A., Natcher, D., & Johnston, M. (2014). Linking gender, climate change, adaptive capacity, and forest-based communities in Canada. *Canadian Journal of Forest Research, 44*(9), 995–1004. https://doi.org/10.1139/cjfr-2014-0174

Rodó-de-Zárate, M., & Baylina, M. (2018). Intersectionality in feminist geographies. *Gender, Place & Culture, 25*, 547–553. https://doi.org/10.1080/0966369X.2018.1453489

Rojas-Downing, M., Pouyan Nejadhashemi, A., Harrigan, T., & Woznicki, S. A. (2017). Climate change and livestock: Impacts, adaptation, and mitigation. *Climate Risk Management, 16*, 145–163. https://doi.org/10.1016/j.crm.2017.02.001

Rubio, A., & Roig, S. (2017). *Impactos, vulnerabilidad y adaptación al cambio climático en los sistemas extensivos de producción ganadera en España.* Oficina Española de Cambio Climático. Ministerio de Agricultura y Pesca, Alimentación y Medio Ambiente.

Sachs, C. E., Barbercheck, M. E., Brasier, K. J., Kiernan, N. E., & Terman, A. R. (2016). *The rise of women farmers and sustainable agriculture.* University of Iowa Press.

Smith, D. E. (1987). *The everyday world as problematic: A feminist sociology.* University of Toronto Press.

Vaccaro, I., & Beltran, O. (2009). Livestock versus "wild beasts": Contradictions in the natural patrimonialization of the Pyrenees. *Geographical Review*, *99*(4), 499–516. https://doi.org/10.1111/j.1931-0846.2009.tb00444.x

Venkatasubramanian, K., & Ramnarain, S. (2018). Gender and adaptation to climate change: Perspectives from a pastoral community in Gujarat, India. *Development and Change*, *49*(6), 1580–1604. https://doi.org/10.1111/dech.12448

Walker, H. M., Reed, M. G., & Fletcher, A. J. (2021). Applying intersectionality to climate hazards: A theoretically informed study of wildfire in northern Saskatchewan. *Climate Policy*, *21*(2) 171–185. https://doi.org/10.1080/14693062.2020.1824892

Whyte, K. P. (2014). Indigenous women, climate change impacts, and collective action. *Hypatia*, *29*(3), 599–616. https://doi.org/10.1111/hypa.12089

Wilmer, H., & Fernández-Giménez, M. E. (2016a). Some years you live like a coyote: Gendered practices of cultural resilience in working rangeland landscapes. *Ambio*, *45*, 363–372. https://doi.org/10.1007/s13280-016-0835-0

Wilmer, H., & Fernández-Giménez, M. E. (2016b). Voices of change: Narratives from ranching women of the Southwestern United States. *Rangeland Ecology & Management*, *69*(2), 150–158. https://doi.org/10.1016/j.rama.2015.10.010

Winker, G., & Degele, N. (2011). Intersectionality as multi-level analysis: Dealing with social inequality. *European Journal of Women's Studies*, *18*(1), 51–66. https://doi.org/10.1177/1350506810386084

# Reflection on Chapter 5

## The Scarlett Attack[1]

*Lucía Cobos*

**Lucía Cobos González** *was born in Cádiz in 1979 and has been wearing glasses since she was 11 years old. She grew up in Seville, where she studied Audiovisual Communication. For 15 years she worked as a journalist and cultural manager, especially in the book sector, both in Seville and Madrid. In 2015 she stepped into livestock farming with her brother and they took over a family farm in Cumbres de Enmedio (Huelva), with the constant support and help of their family. She is the mother of Lola (2017) and Nicolás (2020). Neither of them wears glasses (so far).*

*This reflection has been translated from its original version in Spanish.*

A few years ago, seven, I think, I used to live in downtown Madrid and I had another life. I had another profession, other interests, I had never seen a cow up close; I did not know how long the pregnancy of a sheep lasted. But I was not happy.

Now I have two children, eighty cows, a hundred sheep, I have saved a lamb's life, and I want to learn to do management that protects and regenerates the land of my ancestors. I am also aware in the first person how difficult it is to do things correctly and how the climate and managing a family farm have changed in 30 years.

Others say I was hit by the "Scarlett attack." History tells that this land, *Las Cumbres*,[2] was bought by my great-great-grandmother. Then, it was managed by my great-grandmother. After that, for two generations, the land was distant from us. Now, this land has returned to us, we work it, and we take care of it between my brother and me. Sometimes together and sometimes against each other. We disagree on "authority": He is a veterinarian, I am a journalist; we disagree on rearing issues: Animals and children; on priorities, on negotiations. But we also agree, especially in the common goal of caring for and improving our children's heritage.

In recent years I have learned that there is much more "truth," much more reality, much more life, in the countryside than in the city. However, this return to

DOI: 10.4324/9781003089209-11

the essential also implies fears that from the asphalt [of the city] seem far away. For example, now that I know how to appreciate the seasons, I am more afraid of summer, and I value and appreciate every rain.

## The Heat and the Rain

The heat and the rain mark each season. Naturally, the cattle movements within the farm depend on them, the health of the animals, the number of inputs we need to keep them well fed. But also, the quality of my sleep and the intensity of my sleepless nights depend on the heat and rain.

The feeling of vulnerability in the countryside is permanent. The joys, the health of your animals, and the income of your business depend on many factors that are out of your hands.

The heat and the rain weave together every conversation about the evolution and current situation of livestock.

Heat and rain are always subject to the same phrases: Each year, it is warmer for longer; every year, it rains less.

## These are Pennies[3] That Fall from the Sky

All my life, I heard my grandfather say this sentence while it was raining: "These are pennies that fall from the sky." But I had to start my livestock farmer phase to truly understand what that meant. Water is everything.

In the city, water is also everything. When I lived in the big city, I appreciated it because the rain cleaned the polluted air. Still, in the end, there are always more inconveniences that it causes, for example, traffic jams, umbrellas getting in the way, people more angry than usual. Finally, the thought was: "Another rainy day, how annoying."

But when you get to the countryside, the perspective changes. Without water, you cannot do anything. You cannot raise animals without water. How many litres does a cow drink? And a lactating cow? And the grass? Oh, dear ones, the grass only grows if it rains.

What I did not remember about my grandfather's phrase were the buts. Common quotes are famous for a reason, and that phrase that says that it never rains to everyone's liking could not contain more truth.

Oh, how nice that it is raining. No, now it is not good that it's raining because the pasture gets wet and what remains is spoiled. Good, it's raining. No, now it is not good for it to rain because if it does not rain again in 15 days, the acorns dry up. Good, it's raining. No, now it must not rain because the hay that has already been collected gets wet. No, not now because... no, now not because...

Seriously? I have been learning that for the livestock farmers to be happy, it has to rain on specific dates, at a particular angle, and in exact amounts. But I disagree. Let it rain, please, let it always rain, because without water, there is no hay, no acorns, no cows, and no us. *Carmiñita* the thirsty, I used to tease my mother in the early years. How innocent I was, it is clear that now I am the thirsty one.

I look at the sky in distress. No rains are expected for 15 days, and tomorrow a straw truck arrives. The water levels in the pond continue to drop after the long summer. Any day the subsurface water samples will give us a scare.

In recent years, we have learned to look at contour lines to bring water to the different paddocks. We have to search for different springs, spend a lot of money finding water, and get water to different parts of the farm. Today, livestock management must do controlled grazing, move the cattle through the different areas to take advantage of the pasture and allow it to recover. But that is impossible if they cannot drink everywhere.

My mother and aunt saw the water level drop in the pond two years ago for the first time in their lives. We no longer know what to invent to avoid breakdowns because losing part of a deposit guarantees us days of anguish until we see if it can recover.

The optimal stocking rate does not have to do with how many animals can feed on a farm. Instead, the optimal stocking rate for a farm is set by how many animals can drink on it.

*These are pennies that fall from the sky*, said my grandfather, and every day we are poorer.

## Wolves Do Not Eat the Cold

Wolves do not eat the cold; they told me the first "warm" January. And March came with a cold that made the Easter holidays seem like Christmas.

But what about the heat?

The heat always bothered me a lot, and the years go by, and it only grows. With each passing year, it is warmer and the heat is more prolonged. May, June, July, August, September, October. Seriously? Six months of summer? Five, six months are many months of concern about fires, food and water for animals, many months in which work in the fields has to be adapted to the heat schedules.

Every year we try to delay the lambing season[4] from August–September to avoid flies[5] infesting the ewes. But if we delay it, the lambs do not reach the right weight for the good sales season, before Christmas. But if we lamb early, we have a lot of insect problems, and in the end, the ewes have a hard time, and so do we. "It is that before, the month of September was not so hot," "before...." These are not old-time phrases. I am 41 years old, and I think I still have time to get old. Still, I can perfectly remember the summers and winters of my childhood, much closer to the image we have of them than to what we have today.

There are also discrepancies here. But now that I think about it, almost all differences with men around me are summarized in this: Change reality or adapt to it. Find water in other pastures or bring the animals to where there is water in summer; make a shed so that the ewes that just gave birth are more protected or delay the lambing date. Adaptation, adaptation sometimes understood as resignation, it is true, but also as a form of resistance, of maintaining a minimum of sanity.

Now that I think about it, this is also what happens to me every day at home, with the father of my daughter and my son. And it is that in the end, raising cows, sheep, and children is very similar. Like a vegetable plot. I suspect that is like everything that has to do with nature. They are living beings, and we must adapt to them, adjust our expectations to their realities.

Here's a wonderful can of worms ready to be opened in a world full of men. What farm woman can weigh a truck of straw on the scale at 7:30 in the morning? It is clear, only the one who does not have children.

I started talking about the cold and the heat, and I end talking about the children. Like water and heat, raising children also conditions the management and handling of a farm.

## Where Have the Birds Gone?

The year 2020 came ready to show us our fragility, and in addition to a son and a pandemic, it also gave us a wildfire in *Las Cumbres*.

Torrential rains are very, very scary, but the fire is another level. Since we started with this project, every summer we repeated that an action protocol was necessary in case of a fire. Still, the issue remained there, in limbo, floating among the taboos of the countryside.

FFPP: Forest fire prevention plan. We have a FFPP. We expanded the FFPP. We run the FFPP almost well. But hehe, this is another taboo: Everyone has a FFPP, but hardly anyone executes it well.

And suddenly, on 31 July 2020, the same day we found out that Nico, my second son, had his first tooth coming out: "Run! Your cows are burning."

Ten hours later, the fire was under control, and we had been saved from total disaster thanks to the wind blowing in another direction. During those ten hours, I prayed to all the gods that I know—and look, I like classical mythology—but I did not set foot in the field. It was my turn, once again, to adapt to the circumstances, to stay and take care of my family: My mother, my father, my aunt, my daughter, and my son. Accept that my son was six months old and needed me more than my animals.

Stepping on the earth, your earth, when it has burned, is a feeling that I do not wish on anyone. Silence, the most profound silence I remember, total silence.

Here I do refuse to adapt: What can we do to prevent fires? Grazing, firebreaks, FFPP, none of that is enough. So once again, we return to the heat and the water. Is adaptation all we have left?

## Notes

1 Referring to the protagonist of *Gone with the Wind.*
2 The farm's name.
3 The original Spanish version reads: These are *duros* that fall from the sky. The word *duros* refers to the old 5 pesetas coin.
4 In the original Spanish version, the word is *paridera. Paridera* is a Spanish word that in sheep husbandry refers to the lambing season.
5 Myiasis.

# 6 Leadership in Mountain and Wildland Professions in Canada

## Examining the Impacts of Gender, Safety, and Climate Change

*Rachel Reimer and Christine Eriksen*

**Rachel Reimer** *is a PhD Candidate at the School of Geography and Sustainable Communities, University of Wollongong, Australia. Based in Revelstoke, BC, their work takes an intersectional approach to organizational culture change in mountain-based professions. Rachel has operational experience leading in the field as a wildland fire initial attack crew leader, and as an avalanche educator and guide. She was the Association of Canadian Mountain Guides' Diversity and Inclusion Committee founding chair, and served as co-chair for three years. She is currently co-lead on a Deep Dive into Culture for the US Forest Service, a four-year project facilitating culture change in harassing and discriminating norms.*

**Christine Eriksen** *is Senior Researcher in the Center for Security Studies at ETH Zürich, Switzerland. With a focus on social dimensions of disasters, she gained international research recognition by bringing human geography, social justice, and wildfires into dialogue. Christine is the author of two books and over 75 articles and book chapters, which examine social vulnerability and risk adaptation in the context of environmental history, natural hazards, cultural norms, and political agendas.*

## Introduction

Mountain and wildland environments present seasonally specific hazards that require risk management strategies for people to coexist safely among them. This chapter explores the intersectional gendered dimensions of professional working cultures situated in mountain and wildland environments. It focuses on winter guiding and avalanche work, and summer wildland firefighting in Canada's mountainous western provinces: British Columbia and Alberta. The chapter builds on two case studies. The first case study took place in British Columbia during the 2016 wildfire season (Reimer & Eriksen, 2018). The second was a national case

DOI: 10.4324/9781003089209-12

study conducted during the 2019 avalanche and guiding season (Reimer, 2019). The chapter first explores mountain and wildland environments as both a socially co-constructed masculine space and as a physical space made increasingly complex by climate change and its effects on wildfire and avalanche risk. Second, it describes the use of a feminist appreciative approach to Action Research for in-depth inquiry into the intersectional gendered dimensions of professional mountain and wildland cultures. Finally, it shares the combined findings, conclusions, and significance of the two case studies, including participant-driven recommendations for more effective, inclusive, and safe social interactions amid increasingly complex environmental hazards.

## Gender within Mountain and Wildland Environments

The process of human engagement with the reality of environmental topography and hazards begins with a series of interwoven narratives about both mountains and wildlands, which provide meaning before we even enter the physical space. As conscious beings, we "live in language" and make meaning of our experiences; we "observe each other and observe ourselves through each other" (Efran et al., 2014, p. 8). For many, mountain and wildland environments become "wilderness" through a personal and communal process of meaning-making. Wilderness is both a literal place and a fluid concept: "quite profoundly a human creation" linked to specific cultures and times in history; it is a "mirror" reflecting our "unexamined longings and desires" (Cronon, 1996, p. 7). As part of the occupation of Indigenous lands that became known as Canada, literate Europeans crafted a concept of wilderness that sought "self-discovery," and centred masculine European identities (Sontag, 1966). "The frontier emerges as the space in which the young man proves his manhood and the older man reclaims his youth. This frontier, existing in fantastical masculinity, has no temporality" (Fiske, 2004, p. 62). The space itself is defined by those who interact with it. Physical realities are overlaid with conceptual realities in collective and individual processes of meaning-making. This historic and fictive frontier mentality in Canada's west has shaped today's narratives about who belongs in mountains and wildlands, and who we become in these environments.

In his extensive study of ski culture in British Columbia, Stoddart (2011, p. 109) identified how popular ski media and participants in his study characterized "the mountainous sublime as a site for performing athletic, risk-seeking masculinity." The recreational skiing culture's performative masculinity described by Stoddart mirrors the culture of "rural masculinity" within wildland firefighting (Bye, 2009; Desmond, 2007). Rural masculinity configures remote, non-urban environments as a masculine space, where "the relationship between nature and the [male] body revolve around bravery, fearlessness, toughness, physical fitness, and an ability to disregard discomfort and pain" (Bye, 2009, p. 280; see also Desmond, 2007). The associations between technical mountainous terrain and masculinity in the avalanche and guiding industry have also been established

(Walker & Latosuo, 2016), distilled as a concept into "mountain masculinity" (Reimer, 2019). If non-urban, rural, mountain, and wildland spaces are configured as masculine "frontier" spaces, then the types of people we must supposedly become in order to be worthy of belonging there are linked to performing a specific set of masculine norms and ideals (Eriksen & Waitt, 2016; Eriksen et al., 2016; Reimer & Eriksen, 2018). The processes by which mountain and wildland spaces and human performance in those spaces become gendered is not a static relationship, but a co-creative act in which we participate. It is a "gender-environment nexus" (Nightingale, 2006).

In this chapter, when we speak of gendered spaces and performative gender, we ground our understanding of gender in a fluid and inclusive definition—a definition freed from a specific sexed body or fixed set of attributes, and inclusive of men (hooks, 2000; Kimmel et al., 2005). For our purposes, we define *gender* as a fluid spectrum of behaviours that all people co-create by interacting together in a system, and as a socially negotiated set of variables that are expressed and/or imposed through constantly changing power relations (Eriksen, 2014; Itzin & Newman, 1995; Maxfield et al., 2010; Pacholok, 2013).

In addition to mountain and wildland environments as a gendered performative space, we consider the host of physical hazards that pose additional risks to people who choose to live and work there. The hazards that affect the guiding, avalanche, and/or wildland firefighting professions include dangerous weather and avalanche conditions: A technically difficult (nearing vertical, vertical, or overhanging) mountainous terrain; altitude, snow and ice conditions; risks in accessing these remote places via foot, vehicle, aviation, or otherwise; and the dynamic geological and meteorological nature of both snow and wildfire environments, which make decision making challenging (McClung & Schaerer, 2006; Stewart-Patterson, 2008; Walker & Latosuo, 2016). Furthermore, the increasing frequency, intensity, and uncertainty of many of these hazards with climate change bring added complexity to decision making in mountain environments for both the avalanche forecasting and mountain guiding professions (Ballesteros-Cánovas et al., 2018; Huggel et al., 2012), and for the wildland firefighter profession (Maxwell et al., 2020; Sommers et al., 2014). This presents a challenging environment within which to manage risk, survive, and thrive as professionals and leaders.

Taken together, the complexity of mountain and wildland environments as both performative gendered, and geologically and climatically hazardous spaces provide rich learning and theorizing opportunities. For example, is the performance of gendered cultural norms, as a prerequisite for belonging in mountain and wildland spaces, safe and sustainable? How do performative gendered norms affect the practice of leadership and decision making when teams face complex hazards? What does embodied feminist leadership in the mountain and wildland environment look like? Can we imagine—or perhaps reclaim—these environments as a non-gendered space that fosters the emergence of tolerant and inclusive professional mountain and wildland cultures?

## Case Studies and Methodology

### Background

Both wildland firefighting and the avalanche and guiding industry are remote, wilderness-based work environments operating in small-to-medium-sized teams. The British Columbia Wildfire Service (BCWS) has wildland fire response supervision in 14 different bio-geoclimatic zones province-wide, ranging from low to high fire frequency landscapes covering 94 million hectares of combined private and public land, 75% of which is considered mountainous terrain. Their wildland fire crews (3- or 4-person initial attack or rappel, or parattack, and 20-person unit crews) respond to approximately 2000 wildfires per year (BCWS, 2018).

The avalanche and guiding industry in Canada is comprised of three industry associations: The Association of Canadian Mountain Guides (ACMG); the Canadian Ski Guides Association (CSGA); and the Canadian Avalanche Association (CAA). The ACMG has guiding certifications spanning from indoor climbing gym instructors in urban areas, to hiking, outdoor rock climbing, ski, and alpine climbing, to mountain guides, who are certified to work in the most technically difficult mountainous terrain in remote environments and are able to guide worldwide. The CSGA certifies ski guides to work in helicopter and snow-cat backcountry skiing operations in Canada. The CAA provides training and professional industry-level avalanche forecasting certifications, which are required for any guide working in avalanche terrain (both the CSGA and ACMG guiding certifications require a CAA forecasting certification as a prerequisite); ski patrol; forecasting for mining, resource extraction, or transportation corridors threatened by avalanche hazard. Guides and avalanche workers typically work as part of teams that range in size from 2 to 25 people, though solo work is also acceptable, and often the norm for summer (and some winter) guiding.

### Demographics

The two mountain- and wildland-based professions have workforces that are predominantly White and male. The avalanche and guiding profession in Canada is 95% White and approximately 85% male (CAA, personal communication, May 2019); the BCWS is 75% male (Reimer & Eriksen, 2018). This prevalence of a majority White, male workforce is linked to the "masculinity contest culture" (MCC) commonly found among historically male-dominated occupations. In a study by Glick et al. (2018, p. 462), MCC was used as a scale to assess workplace cultures against four cultural dimensions: "show no weakness; strength and stamina; put work first; and dog eat dog." Participants who perceived their workplaces to be high in these masculinity contest cultural norms experienced their workplaces as "dysfunctional, rife with negative social behaviours, dissatisfying, and personally harmful" (Glick et al., p. 466). In the following findings and discussion sections, we discuss the implications of a MCC in greater depth. It is worth noting here that male-dominated professions also tend to be relatively homogenous due to the underrepresentation of historically oppressed Peoples in the workforce, such as Black, Indigenous, and People of Colour (BIPOC) and LGBTQ2SIA+ people.

In another BCWS study, Pacholok (2013, p. 40) highlighted that traditionally male-dominated "masculine" jobs are "most readily filled by a person who has similar characteristics to those already working in the occupation–in the case of [firefighters], primarily white, heterosexual, 'masculine' men." Walker and Latosuo (2016) similarly point to the "historical lack" of traditionally oppressed Peoples in mountaineering, noting that women climbers on Mt. Denali between 1999 and 2012 accounted for 13% of total climbers on the peak. New or aspiring entrants into these workforces whose identities align with the identities of current leaders (either appointed leaders or leaders via social power) are more likely to be perceived as a "good fit" within the workforce, while the opposite is also true (Jackson & Parry, 2011). Is it all about belonging and (in)validation? Or are there other factors that create "natural" non-discriminatory barriers to employment for smaller-statured, less "masculine" persons, given the physical demands of both wildland firefighting and mountaineering?

### Physical Fitness for Mountain and Wildland Environments and Professions

The question of physiological barriers to entry and success in mountain-based professions has been explored and debunked by research on physiological limitations to performance in mountain climbing at high altitudes (Bhaumik et al., 2004; Huey et al., 2007). Firefighting fitness test trials also show that females' lower aerobic capacity is not a limiting factor for success at strenuous task completion: "as expected women reached lower VO2max, had higher physical strain, and performed more poorly than men in several of the...tests. However, on all tests some women performed better than some men" (Lindberg et al., 2013, p. 6). In 1999, a British Columbia Supreme Court ruling recognized the "duty to accommodate" in fitness testing, after a female firefighter was fired for falling 50 seconds short of the 11-minute cut-off during a timed running test after effectively performing all the strength-based testing. The Supreme Court cited that firefighting task performance was not simply a measure of aerobic output, but rather a combination of skillsets that included, but was not limited to, aerobic capacity (Meiorin v. Province of British Columbia, n.d.). The lack of evidence for physically strenuous task performance as a reliable justification for excluding women as a biological "other" from the workforce, makes it plausible that smaller statured men and people from non-White, non-heterosexual, non-male identities might also experience their physical abilities as sufficient for working in mountain and wildland professions despite narratives claiming otherwise. If biophysicality is not a significant factor, is it the mental strength required to face the hazardous environment, which results in the wildland firefighting and guiding and avalanche professions remaining almost exclusively White, and mostly male?

## Risk Management and Gender

In mountain and wildland environments, physical hazards related to technical and steep terrain, dynamic conditions, and uncertainty about changing conditions, work together to create a highly complex decision-making environment for avalanche and guiding professionals (McClung & Schaerer, 2006; Stewart-Patterson, 2008) and wildland firefighters (Barton et al., 2015). This poses a significant *internal* challenge. Traditionally, men have been assumed to have greater capacity for managing and tolerating risk than women (Hohnisch et al., 2014; Maxfield et al., 2010; Sarin & Wieland, 2016). There is some evidence that aversion to engaging in risk management is a learned behaviour vis-à-vis feeling uninvited into the risk-based conversations, for example, in wildfire risk management (Eriksen, 2014, 2019). During a 2014 climbing season on Mt. Denali, researchers conducted a gender-based study of mountain guides' risk management and decision making (Walker & Latosuo, 2016). The study found that female risk-taking was not measurably different from their male counterparts, and male versus female risk tolerance was not statistically significant, yet female decision making was perceived to be more conservative than that of male counterparts.

Perception of risk-taking relies on the visibility of risk-taking. The question of visibility of risk-taking behaviours amongst genders was explored by Maxfield et al. (2010). Their findings reveal that female risk-taking is "hidden" as an adaptive mechanism to avoid hyper-criticism for perceived risk management failures amongst female risk-takers. The visibility of female leaders, and the negative associations they experience with being seen as risk-takers, mean some women "hide" as an adaptive strategy to avoid criticism. Wielend (2012) compared decisions involving risk where potential outcomes are known (e.g., winning or losing a bet), with decisions involving risk where uncertainty is high and potential outcomes are unknown (e.g., skiing an avalanche slope; reconnaissance on a wildfire). In risk-based decisions with high uncertainty, gender did not affect how decision-makers behaved. Rather, in navigating risk with uncertainty, the decision-maker must "rely on their own, internal, subjective expectancies of the probabilities of outcomes" (Wieland, 2012, p. iii). The mental fortitude required to manage risk in mountain and wildland environments is therefore not exclusively a "masculine" characteristic. Risk management is, in fact, a realm where women perform on par with men in terms of measurable outcomes.

Given the mountain and wildland professions' demographic make-up as traditionally White and male-dominated, and the lack of biophysical or decision-making limitations on mental capacity to performance that would otherwise prevent people of other identities from inclusion in the profession, we turn now to an inquiry into the gendered culture at the heart of these professions.

## Inquiry Methodology

Both inquiries used Action Research (AR) methods, employing an adaptation to the traditional three-part cycle by using the Action Research Engagement (ARE) model (Rowe et al., 2013). AR prioritizes practical outputs to help transform challenges within organizational culture (Stringer, 2014). It is a collaborative process of complex problem-solving designed to foster investment in creating organizational change (Coghlan & Brannick, 2005; Reitsma-Street & Brown, 2004). AR and ARE, evolved out of an organizational-based approach to creating collaborative change, while other participatory approaches have their roots in community or education paradigms (Thiollent, 2011). The ARE process provides a "look" and "think" phase to AR, and makes recommendations for the third "act" phase of traditional AR, without implementing them. This approach provided space for sensitivity to navigate the fear amongst organizational leaders when critically approaching gendered cultural norms. It also supported us as researchers in incremental ways to ensure project success by managing documented resistance to change within traditionally male-dominated mountain- and wildland-based professions (Eriksen, 2014; Pacholok, 2013; Reimer & Eriksen, 2018).

We also took a feminist appreciative approach to ARE—a robust combination that provided us with the most appropriate framework within which to work. Appreciative Inquiry (AI) is a five-part process that focuses on using positive future-oriented "ideals" to mobilize group resources in an imaginative and strengths-based approach to change. AI seeks to imagine the best of "what is," and imagine what could be (Bushe, 1998, 2012). This study did not utilize the full AI process, but combined the values of AI with a feminist approach. A feminist approach is sensitive to the power relations that arise, and to systems of domination in a hierarchical, White, heterosexual, male-dominated professional culture with known gendered norms (Hesse-Biber, 2012; Taylor, 1998). We relied on hooks' (2000, p. viii) definition of feminism, which is "a movement to end sexism, sexist exploitation, and oppression." This allowed us to focus on systems of domination as the "problem" rather than the people involved. We found a feminist appreciative approach to ARE to be effective in supporting inquiries into the complex, intersectional, gendered cultural norms in male-dominated mountain- and wildland-based professions, especially methods with the goal of creating sustainable organizational cultural change.

### Research Ethics, Scope, and Limitations

Both studies followed the research guidelines established by the Canadian Tri-Council Policy Statement for research (Canadian Institutes of Health Research [CIHR] et al., 2014). The BCWS study took place during the 2016 fire season, after review by Royal Roads University's research ethics committee. The avalanche and guiding study took place during the winter of 2019 after being commissioned by industry associations, and followed the Canadian Tri-Council Policy Statement guidelines.

In both studies, participation was limited to people who, at the time of the research, were either employed by the BCWS (Reimer & Eriksen, 2018) or active members of one of three avalanche and guiding professional associations (Reimer, 2019). This limitation meant that people who had left their employment were not included, resulting in a lack of insights from people who might have been pushed out of their profession due to discrimination, harassment, or other reasons. Similarly, aspiring wildland firefighters, guides, and avalanche workers were not included in the research. Given the topic of research and the masculine norms within the industry, some White males may have self-excluded from participation due to a perceived threat to their identity (Schwalbe & Wolkomir, 2001), or due to a perception of being exempt from the topics (Kimmel et al., 2005). Similarly, some women in traditionally male-dominated professions create a workplace persona of toughness, which may in turn cause them to view open discussions related to gender identity as a threat (Eriksen, 2019; Reimer & Eriksen, 2018).

### *Data Collection*

Both studies build on the notion that the "wisdom of the crowds" delivers more intelligent solutions to shared problems (Surowiecki, 2004). Each aimed to engage multiple standpoints to arrive at a more reliable and valid understanding of reality (Golafshani, 2003). This was achieved in the BCWS study by using two methods of data collection: An online conversation tool called ThoughtExchange sent from the employer to all staff via government email (n = 205) and semi-structured interviews, solicited via the same method as the survey (n = 5). ThoughtExchange allowed participants to share their original thoughts in responses to questions, and also to view other participants' thoughts and rank them by assigning stars to thoughts that resonated with them. This facilitated an anonymous, online conversation that democratized findings and provided an opportunity for cultural norms to be observed by all participants.

In the avalanche and guiding study, it was achieved by using an online survey (n = 514) that was inclusive of all three industry associations, on a spectrum from "industrial" resource extraction and transportation avalanche work to "service-oriented" guiding work. It is notable that this industry study was the first time the three industry associations had collaborated to reach a shared goal. In the BCWS study an inquiry team comprising four wildland firefighters was used to pilot questions, provide feedback during analysis, and enhance critical self-awareness for the lead author. In the avalanche and guiding study, questions were piloted by industry association leaders and four other individuals in the profession, and an external researcher was hired to provide feedback during analysis, and to enhance critical self-awareness. Inviting these collaborative and critical forms of feedback into the research process enhanced validity and reflexivity (Gergen, 2000; Hesse-Biber, 2012).

### Findings

In the following discussion, we focus on three key findings that emerged in both studies as being significant for understanding the gendered dimensions of mountain- and wildland-based professional working cultures (for in-depth details of all results, see Reimer, 2019; Reimer & Eriksen, 2018). These findings highlight how:

1) Performative masculinity in the selected mountain and wildland professional spaces is linked with perceptions of competence;
2) Espoused values of diversity and inclusion are misaligned with exclusionary cultural norms in practice;
3) Participant-based recommendations for cultural change strategies indicate a need for critical dialogue and increased self-awareness.

### *Performative Masculinity in Mountain and Wildland Professional Spaces Is Linked with Perceptions of Competence*

In the BCWS study, the strongest opinion cited by participants was that gender—defined within the ThoughtExchange online tool as "the attitudes, feelings, and behaviours that a given culture associates with a person's chosen or biological sex" (Reimer, 2017)—makes a difference to how wildland firefighters are treated at work. Survey participants inductively shared 51 responses that linked skills performance, perception of competence, and gender. For example, "Women are often belittled or have their abilities questioned"; "men are presumed competent whereas women need to initially prove competence." These behaviours were especially linked to "physical or tactical leadership roles." Four out of five interview participants agreed with this linkage. One interviewee described how performing a gender other than the dominant (cis-male) gender was "super lonely [because] you notice the difference and you feel kind of awkward and you don't really want to be there." ThoughtExchange participants and interviewees shared the pain of not being chosen for tasks because of assumptions of incompetence or lack of initiative. These exclusionary behaviours were further described as passive aggressive, subtle, and hard to define. Some responses blamed female firefighters for these experiences: "We hire women who consistently underperform"; another ThoughtExchange participant explained that women experience sexism "because you're terrible at your job." One interviewee explained how the blaming of non-dominant genders for their own experiences of discrimination created a lose-lose scenario because if one complained about discriminatory behaviour, then complaining was seen as proof of incompetency and weakness.

In the avalanche and guiding study, the culture of the profession was most frequently described by participants as "exclusive," with an "ego-driven, not self-aware, male-dominated, bro culture" with "bro," "boys club," or "male-dominated" behaviours linked to perceptions of competence. When asked if gender made a difference in how people are treated in the profession, 59.9% said yes, and 30.8% said no. Participants explained, "as a White male, I am the sought after 'ideal' of a guide and I am treated as such" and "if you are not White male, you better be ultra-competent and strong." One participant described the effects of internalized sexism in their workplace:

> There is a huge assumption of competence if you are male. The opposite is present if you are female at our workplace. I have personally heard the one more senior female state that the only reason the guys like her at work is because she hardly talks.

Performative masculinity and competence were strongly linked, with impacts for all genders. In both datasets, mountain- and wildland-based professionals identified a positive association between competence and hypermasculinity, and a negative association between competence and other expressions of gender. Traditional gender roles exacerbated the discriminatory behaviours. In the BCWS study, expressing emotion was seen as "weak" and to be avoided, especially for men. One interviewee shared how "men are pretending to be something that they're not…it's like men are always trying to be an alpha male even if they're not an alpha male." This kind of masculinity also revealed itself in the avalanche and guiding study. Male workers were expected to "buck up and hide their emotional trauma" whereas female workers could "show their emotions." Women were expected to be "soft," "emotional," and take on less physically strenuous tasks, while men were expected to be "hard," "unemotional," and take on more difficult tasks with greater risk. This was especially the case for guides and avalanche workers who were also mothers, who described being treated as if motherhood compromised their risk management abilities and competence.

In addition, the practice of performative masculinity as a barrier to mutual accountability surfaced in both datasets. A firefighter shared, "When you have a bunch of buddies backing each other…no one's going to speak up even if they're a witness to something." One avalanche and guiding professional shared: "I've been told by a male co-worker that my breasts are why [clients] like me. I've sat in a guides' meeting where my physical attributes were discussed. I've been groped by male guides and clients." The firefighting imagery of a "bunch of buddies" being complicit in harmful behaviours is mirrored in the experiences of guide and avalanche workers, with openly derogatory and shaming events going unchallenged by the group. This included transphobic and homophobic statements. One individual shared "I am transgender. I am not 'out' to my coworkers. I fear that I would not be treated equally due to the comments and jokes I hear on a daily basis."

The avalanche and guiding study dug deeper into quantifying experiences of discrimination and harassment (Canadian Human Rights Commission, n.d.; Carleton University, 2019). In a self-reporting question, 46% of female and 3.5% of male respondents reported experiencing gender discrimination, with the most commonly cited experiences of gender discrimination being (in order of frequency): 1) competence, 2) motherhood, 3) traditional gender roles, and 4) hostile, sexualized work environments. In a self-reporting question on harassment, 27% of females reported experiences of sexual harassment, compared with 4% of males. Of the female experiences, one-third were unwanted touching. There were nine incidents of sexual harassment that were initiated by a supervisor, mentor, examiner (in a guides' exam), or instructor, across all industry associations.

Sexual harassment from clients made up 40% of harassing behaviours reported in this study, with the remaining 60% coming from fellow colleagues.

The next finding discusses the degree to which these behaviours and beliefs are conscious and highlights the cognitive dissonance at play in these professional mountain and wildland cultures.

## *Espoused Values of Diversity and Inclusion Are Misaligned with Exclusionary Cultural Norms in Practice*

In the BCWS study, participants shared their values and ideals in response to questions inquiring into what constitutes "good" leadership, and what an ideal future may look like for them. Both sets of responses pointed to "espoused values" (Schein, 2010) that framed femininity and masculinity in leadership as equally valuable. This meant leaders cultivated an "understanding of their unique attributes, leading to a self-concept that integrate[d] both their strengths and weaknesses" (Reimer & Eriksen, 2018, p. 720). When participants ranked leadership qualities, they most strongly valued "supportive," "respectful," and "humble." In addition, when asked about future growth opportunities, firefighters linked diversity with excellence: It "makes for better decisions," "provide[s] the best product I can put out in my crew," and the future of leadership is one where leaders "facilitate."

In contrast to this rather rosy picture of diversity, excellence, and gender-balanced leadership, firefighters also shared deeply dissonant experiences in the qualitative comments linked to these two questions. One interviewee explained, "Really strong females [in leadership make] the work environment always better" but they are not going to "get on top of the tailgate doing that 'rah rah' speech, follow me into battle…and that's what we perceive to be a leader." This dissonance was further expressed in the trade-off facing female leaders who may embody the espoused values, but in practice are treated as "less than" their male peers. If a female leader took a "feminine role or spin on leadership you're not respected as a leader, but you're [seen as] a 'really great person'." This trade-off between being respected and being liked was due to conforming with feminine gender norms but failing to perform masculinity in leadership (which is conflated with competence, as discussed in the previous section). However, if a female leader performed "competently" in a hypermasculine way, they would be "really respected as a leader" yet seen as "maybe not a nice person"—in this case, a respected/disliked trade-off. This dissonance between espoused and lived values also affected male participants and their practice of leadership. One interviewee explained how they see a "lot of uncomfortable men in leadership who probably if they had support in being who they actually are…might be better leaders."

In the avalanche and guiding study, when asked if the profession was inclusive, 68.5% said yes, 23% said no, and 5% said both (for inclusivity definition see Canadian Task Force on Diversity and Inclusion, 2017). This perception of inclusivity was explained in different ways: "We are quite similar people in general so it's pretty easy to be inclusive to like-minded people" or alternatively:

> They think they are [inclusive] because they believe themselves to be open and easy going but due to many barriers to entry, there are not a lot of non-white, non-male professionals in our industry. If there were, and these places were faced with more diverse resumes, I think they would have trouble putting their bias aside.

This question, when compared with how culture was described by the majority of participants ("exclusive"), points to a misalignment of espoused values and actions within the culture. In addition to "exclusive," participants identified cultural values of "professionalism" and "safety-oriented" as second- and third-ranked in descriptions of the professional culture. On the surface, these seem like common-sense rallying points within a profession dedicated to managing risk and maintaining safety for clients in the mountain environment (ACMG, 2020). However, in explaining their perspectives, participants shared that these values were perceived as neutral and positive by some, and as exclusionary and negative by others. Males aged 45+ tended to focus on a neutral/rational interpretation of professionalism and safety. Others, notably females and males under the age of 45 commented, "safety is used as an excuse to retain male dominance," to be seen in context of "a very male-oriented culture that has significant focus on maintaining its overtly masculine characteristics." These nuances indicate that professionalism and safety are complex, and in some ways gendered. They can, in practice, promote hypermasculinity to the exclusion of other forms of "safe" and "professional" behaviours that are non-masculine.

The dissonance that surfaced in both studies points to a significant lack of consciousness surrounding espoused values and exclusionary behaviours. Espoused values in both mountain professions, on the surface, seem to represent a high-performing and inclusive ideal. Yet in both datasets, the experiences of mountain professionals reveal an entirely different day-to-day reality within the culture, one rife with discriminatory and harassing behaviours that escape meaningful accountability. This presents challenges as we look to a future that is ideally more inclusive than the current realities in mountain and wildland environments.

### *Participant-Based Recommendations for Cultural Change Strategies Indicate a Need for Critical Dialogue and Increased Self-Awareness*

In sharing their future-oriented ideals for the organization, the BCWS study was split equally between desires for a culture where "gender doesn't matter" and is no longer a factor in skills performance, and a future where genders are "equal" and "all crews have diverse representation." Participants suggested action steps to achieve this that emerged in three equal-strength themes within the data: 1) create a conversation about gender in wildland fire; 2) examine hiring and succession planning to ensure a bias-free process; and 3) actively support females in leadership. Firefighters identified the process of creating cultural norms as the target of change actions, rather than focusing on blaming one demographic (e.g., men or women). Complexity arose when discussing actively supporting females in leadership. For example, two interview participants stated that female leaders are "less approachable" than male leaders, or "aren't very supportive of the women that are working below them." These types of behaviours were linked to firefighters needing to perform masculinity in order to be seen as "good" leaders. Participants voiced concerns about supporting females and other minorities without concurrently tackling the gendered cultural norms. Support for traditionally oppressed Peoples must also engage with, and challenge, the belief that they are to blame for their own experiences of discrimination, and must disrupt the association between female bodies, femininity, and poor work performance.

In the avalanche and guiding study, participants' top-three ranked future goals for the profession were: Equality of opportunity, an inclusive and diverse culture, and a respectful work environment. The action steps identified to achieve these outcomes were: 1) more active mentorship and leadership from amongst traditionally oppressed Peoples within the profession; 2) active engagement from employers and industry associations via awareness raising, education, and policy change; and 3) a generational shift within industry leadership. Some blame was placed on older generations within the profession, even though all demographics agreed that greater representation of diverse identities and perspectives in leadership would benefit the profession as a whole. Male guides and avalanche workers tended to be more critical of their male peers, and rely on top-down action as a solution to the problematic cultural norms.

In both studies, participants self-diagnosed their professional culture as lacking in self-awareness around issues related to inclusion, and recommended a more critical engagement through dialogue and collaboration to foster a much-needed increase in awareness. The presence of blame factored more strongly in the avalanche and guiding data set (towards older generations in leadership). In both studies participants indicated a preference for traditionally oppressed groups (women, BIPOC, LGBTQ2SIA+persons) to "mentor" or "lead" towards a more just and inclusive future in mountain and wildland professions. The implications these findings are discussed next.

## Discussion

The existence, prevalence, and embedded nature of performative masculinity in mountain and wildland professions is clear from our case studies. Consciousness of these behaviours, and awareness of the dissonance between espoused values and exclusionary cultural practices, is low. There is a desire for change and an awareness that change begins by critically engaging with cultural norms.

What is the impact of the status quo with its performance of gendered norms as a prerequisite for belonging in mountain and wildland environments? Berdahl et al. (2018, p. 424) argue that the contest to "prove" one's masculinity turns workplace culture into a "zero-sum competition played according to rules defined by masculine norms (e.g., displaying strength, showing no weakness or doubt)." When masculinity is linked with competence, as in mountain professions, then the only way to "win" at the zero-sum game is seemingly to prove competence by performing masculinity (Berdahl et al., 2018; Glick et al., 2018; Reimer & Eriksen, 2018). This leads to a narrow definition of competence-as-masculinity that is socially attained. The dependence on others' perception of one's competence makes the experience of being perceived as competent vulnerable to being lost and prone to ongoing defence (Berdahl et al., 2018). The process of dominance and marginalization occurs simultaneously, in a hierarchical flow of power relations known as "hegemony" (Eriksen & Waitt, 2016; Kimmel et al., 2005). This hegemonic dominance necessarily marginalizes other ways of being, such as alternative forms of competence that are not associated with hypermasculinity.

This manifestation of gendered power relations is problematic. There is strong evidence that when masculinity is associated with competence and becomes the workplace cultural norm, high rates of harassment, discrimination, poor mental health, poor leadership, and ineffectual task delivery occur (Berdahl et al., 2018; Glick et al., 2018; Reimer, 2019; Reimer & Eriksen, 2018). The continuation of this gendered status quo within mountain and wildland environments is unsustainable, even though for some (notably hetero cis-males) who benefit, it is seemingly "working" (Kimmel et al., 2005; Stoddart, 2011).

The conflation of performative masculinity with competence is not safe. Given that risk management is a significant aspect of the work mountain and wildland professionals undertake (McClung & Schaerer, 2006; Stewart-Patterson, 2008), there is cause for concern. Walker and Latosuo (2016) raised the alarm in connection with the perceptions of female guides as being risk-averse not matching their actual risk-managing decision making. They describe it as a "gender heuristic trap." Their analysis indicated that if women are seen as the conservative voice, and a counterbalance to high risk tolerance amongst men, these assumptions could lead to poor risk management among teams. Others have raised the question of safety, experience, and decision making along gendered lines in the avalanche and guiding field as areas worthy of further study (Adams, 2005; Stewart-Patterson, 2008). In wildland fire, a U.S. Forest Service fatality in California prompted an investigation that revealed how gendered norms connecting action with competence and inaction with perceived failure were significant factors in team decision making that resulted in loss of life (USFS, 2016). The connection of performative masculinity and competence with doing, achieving, and proving, as we have outlined above, creates a reinforcing escalation of risk. In the context of climate change that exponentially increases the environmental hazards faced by wildland firefighters, guides, and avalanche workers, women will disproportionately bear the costs if their risk-management decisions are not understood and are subject to stereotyping.

Performative gendered norms affect the practice of leadership and decision making when teams face complex mountain and wildland hazards, as risk-taking has emerged as a means for people who are performing masculinity to "prove" their competence, meaning the more risk one takes, the more masculine, and therefore competent, one is perceived to be (Berdahl et al., 2018; Glick et al., 2018). This creates conditions for an escalation of risk and a marginalization of risk-management alternatives. An alternative may be rejected by the team not based on its effectiveness, but rather based on its perceived value in affirming the team's identity as "competent" vis-à-vis performative masculinity.

# Conclusions

Mountain and wildland environments are socially constructed as a masculine space for people who perform a specific kind of masculinity. Perceptions of competence are connected to performative masculinity, which marginalizes alternatives, including feminine leadership, as "weak" or less-than. The case studies discussed in this chapter reveal the damaging effects the conflation of competence with masculinity has on team decision making, safety, and inclusivity in mountain and wildland environments. Participants identified a critical need for change via an examination of cultural norms and dialogue that bear witness to the hidden costs.

Can we imagine, or perhaps even reclaim, mountain and wildland environments as a non-gendered space that fosters the emergence of more tolerant and inclusive professional mountain and wildland cultures, for example, via embodied feminist leadership? Consciously choosing to subvert the masculine norms in mountain- and wildland-based cultures is an act of deviance, a contravention of social rules that reward masculinity as competence and marginalize alternatives as weak. Deviance is more likely to be associated with negative, socially destructive behaviour than with social change. Yet, *positive deviance* from the status quo is often the first step on the path to innovation (Pascale et al., 2010). The cost of creating change is higher for gender minorities in a group (Brown et al., 2015). However, those who are already fighting for respect and inclusion of their identities, in a culture that gives them little space to breathe, are often the most motivated to work for change. Embodied feminist leadership in mountain and wildland environments may critically interrogate the status quo (hooks, 2000), or it may be a quiet internal cultivation of self-worth and self-confidence. More research is needed to inquire into the strategies that women, Black, Indigenous, and People of Colour, and LGBTQ2SIA+ use to deviate, innovate, and thrive on their own terms as professionals in mountain and wildland environments.

Ultimately, our aim with this chapter is to bear witness to the problematic intertwining of masculinity with competence in mountain and wildland environments. It is not meant to intimidate, deter, or harm those performing non-masculine gender identities. Nowhere in our work has concrete evidence emerged to support the theory that mountain and wildland environments and professions are exclusively suited for cis-males or people who perform a certain type of masculinity. In tracing the making of that narrative, we hope to reveal it for the cultural myth that it is. In so doing, we simultaneously bear witness to the spaces and possibilities that are created, when recognizing that human potential in mountain and wildland environments is never truly limited.

# References

Adams, L. (2005). *A systems approach to human factors and expert decision-making within the Canadian avalanche phenomena* [MALT Thesis, Royal Roads University, Victoria, BC, 284]. https://citeseerx.ist.psu.edu/viewdoc/download?doi=10.1.1.471.5010&rep=rep1&type=pdf

Association of Canadian Mountain Guides. (2020). *Who we are.* https://acmg.ca/03public/about/who.aspx

Ballesteros-Cánovas, J., Trappmann, D., Madrigal-González, J., Stoffel, M., & Eckert, N. (2018). Climate warming enhances snow avalanche risk in the Western Himalaya. *Proceedings of the National Academy of Sciences of the USA, 115*(13), 3410–3415. https://doi.org/10.1073/pnas.1716913115

Barton, M. A., Sutcliffe, K. M., Vogus, T. J., & DeWitt, T. (2015). Performing under uncertainty: Contextualized engagement in wildland firefighting. *Journal of Contingencies & Crisis Management, 23*(2), 74–83. https://doi.org/10.1111/1468-5973.12076

Berdahl, J., Cooper, M., Glick, P., Livingston, R., & Williams, J. (2018). Work as a masculinity contest. *Journal of Social Issues, 74*(3), 422–448. https://doi.org/10.1111/josi.12289

Bhaumik, G., Sharma, R., Dass, D., Lama, H., Chauhan, S., Verma, S., Selvamurthy, W., & Banerjee, P. (2004). Hypoxic ventilatory response changes of men and women 6 to 7 days after climbing from 2100m to 4350m altitude and after descent. *High Altitude Medicine and Biology, 4*(3), 341–348. https://doi.org/10.1089/152702903769192296

British Columbia Wildfire Service. (2018). *Wildfire service.* www.bcwildfire.ca

Brown, C. M., Olkhov, Y. M., Bailey, V. S., Daniels, E. R. (2015). Gender differences in the social cost of affective deviance. *The Journal of Social Psychology, 155,* 535–540. https://doi.org/10.1080/00224545.2015.1018859

Bushe, G. (1998). Appreciative inquiry with teams. *Organization Development Journal, 16,* 41.

Bushe, G. (2012). Feature choice by Gervase Bushe foundations of appreciative inquiry: History, criticism and potential. *AI Practitioner, 14*(1), 8–20.

Bye, L. (2009). 'How to be a rural man': Young men's performances and negotiations of rural masculinities. *Journal of Rural Studies, 25*(3), 278–288. https://doi.org/10.1016/j.jrurstud.2009.03.002

Canadian Human Rights Commission. (n.d.). *What is discrimination?* https://www.chrc-ccdp.gc.ca/en/about-human-rights/what-discrimination

Canadian Institutes of Health Research, Natural Sciences and Engineering Research Council of Canada, & Social Sciences and Humanities Research Council of Canada. (2014). *Tri-council policy statement: Ethical conduct for research involving humans.* http://www.pre.ethics.gc.ca/pdf/eng/tcps2-2014/ TCPS_2_FINAL_Web.pdf

Canadian Task Force on Diversity and Inclusion. (2017). *Building a diverse and inclusive public service: Final report of the joint union/management task force on diversity and inclusion.* https://www.canada.ca/en/treasury-board-secretariat/corporate/reports/building-diverse-inclusive-public-service-final-report-joint-union-management-task-force-diversity-inclusion.html

Carleton University. (2019). *Getting the facts.* Sexual Assault Support Services. https://carleton.ca/sexual-violence-support/what-is-sexual- assault/getting-the-facts/

Coghlan, D., & Brannick, T. (2005). *Doing action research in your own organization* (2nd ed.). SAGE.

Cronon, W. (1996). The trouble with wilderness; Or, getting back to the wrong nature. *Environmental History, 1*(1), 7–28.

Desmond, M. (2007). *On the fireline: Living and dying with wildland firefighters.* University of Chicago Press.

Efran, J., McNamee, S., Warren, B., & Raskin, J. (2014). Personal construct psychology, radical constructivism, and social constructionism: A dialogue. *Journal of Constructivist Psychology, 27*(1), 1–13. https://doi.org/10.1080/10720537.2014.850367

Eriksen, C. (2014). *Gender and wildfire: Landscapes of uncertainty*. Routledge.

Eriksen, C. (2019). Negotiating adversity with humour: A case study of wildland firefighter women. *Political Geography*, *68*, 139–145. https://doi.org/10.1016/j.polgeo.2018.08.001

Eriksen, C., & Waitt, G. (2016). Men, masculinities and wildfire: Embodied resistance and rupture. In E. Enarson & B. Pease (Eds.), *Men, masculinities and disaster* (pp. 69–80). Routledge.

Eriksen, C., Waitt, G., & Wilkinson, C. (2016). Gendered dynamics of wildland firefighting in Australia. *Society and Natural Resources*, *29*(11), 1296–1310. https://doi.org/10.1080/08941920.2016.1171938

Fiske, J. (2004). And the young man did go north (unfortunately): Reflections on issues in gender and the academy. In M. Hessing, R. Raglon, & C. Sandilands (Eds.), *This elusive land: Women and the Canadian environment* (pp. 57–75). University of British Columbia Press.

Gergen, M. (2000). *Feminist reconstructions in psychology: Narrative, gender, and performance*. SAGE Publications.

Glick, P., Berdahl, J., & Alonso, N. (2018). Development and validation of the masculinity contest culture scale. *Journal of Social Issues*, *74*(3), 449–476. https://doi.org/10.1111/josi.12280

Golafshani, N. (2003). Understanding reliability and validity in qualitative research. *The Qualitative Report*, *8*(4), 597–606. https://doi.org/10.46743/2160-3715/2003.1870

Hesse-Biber, S. (2012). *Handbook of feminist research: Theory and praxis*. SAGE.

Hohnisch, M., Pittnauer, S., Selten, R., Pfingsten, A., & Eraßmy, J. (2014). Gender differences in decisions under profound uncertainty are non-robust to the availability of information on equally informed others' decisions. *Journal of Economic Behavior and Organization*, *108*, 40–58. https://doi.org/10.1016/j.jebo.2014.07.011

hooks, B. (2000). *Feminism is for everybody: Passionate politics*. Pluto Press.

Huey, R., Salisbury, R., Wang, J., & Mao, M. (2007). Effects of age and gender on success and death of mountaineers on Mount Everest. *Biology Letters*, *3*(5), 498–500. https://doi.org/10.1098/rsbl.2007.0317

Huggel, C., Clague, J., & Korup, O. (2012). Is climate change responsible for changing landslide activity in high mountains? *Earth Surface Processes and Landforms*, *37*(1), 77–91. https://doi.org/10.1002/esp.2223

Itzin, C., & Newman, J. (1995). *Gender, culture and organizational change: Putting theory into practice*. Psychology Press.

Jackson, B., & Parry, K. (2011). *A very short fairly interesting and reasonably cheap book about studying leadership*. SAGE Publications.

Kimmel, M., Hearn, J., & Connell, R. (2005). *Handbook of studies on men and masculinities*. SAGE Publications.

Lindberg, A., Oksa, J., Gavhed, D., & Malm, C. (2013). Field tests for evaluating the aerobic work capacity of firefighters. *PLoS ONE*, *8*(7), e68047. https://doi.org/10.1371/journal.pone.0068047

Maxfield, S., Shapiro, M., Gupta, V., & Hass, S. (2010). Gender and risk: Women, risk taking and risk aversion. *Gender in Management*, *25*(7), 586–604.

Maxwell, C., Serra-Diaz, J., Scheller, R., Thompson, J., & Mori, A. (2020). Co-designed management scenarios shape the responses of seasonally dry forests to changing climate and fire regimes. *Journal of Applied Ecology*, *57*(7), 1328–1340. https://doi.org/10.1111/1365-2664.13630

McClung, D., & Schaerer, P. (2006). *The avalanche handbook*. The Mountaineers Books.

Meiorin v. Province of British Columbia. (n.d.). *"Not reasonably necessary": Aerobic fitness test held discriminatory in B.C. woman firefighter victory.* https://www.ehlaw.ca/jan00-bcgseu/

Nightingale, A. (2006). The nature of gender: Work, gender, and environment. *Environment and Planning D: Society and Space, 24*(2), 165–185. https://doi.org/10.1068/d01k

Pacholok, S. (2013). *Into the fire: Disaster and the remaking of gender.* University of Toronto Press.

Pascale, R. Sternin, J., & Sternin, M. (2010). *The power of positive deviance: How unlikely innovators solve the world's toughest problems.* Harvard Business Press.

Reimer, R. (2017). *The wildfire within: Firefighter perspectives on gender and leadership in wildland fire* [M.A., Royal Roads University, Canada]. Retrieved December 1, 2021, from http://www.proquest.com/docview/1887130507/abstract/8140F11FC9694F5EPQ/1

Reimer, R. (2019). *Diversity, inclusion and mental health in the avalanche and guiding industry in Canada* (pp. 1–54). Association of Canadian Mountain Guides. https://acmg.ca/05pdf/2019DiversityInclusionandMentalHealthStudyFINALReport.pdf

Reimer, R., & Eriksen, C. (2018). The wildfire within: Gender, leadership and wildland fire culture. *International Journal of Wildland Fire, 27*(11), 715–726. https://doi.org/10.1071/WF17150

Reitsma-Street, M., & Brown, L. (2004). Community action research. In W. Carroll (Ed.), *Critical strategies for social research* (pp. 303–319). Canadian Scholars' Press.

Rowe, W., Graf, M., Agger-Gupta, N., Piggot-Irvine, E., & Harris, B. (2013). *Action research engagement: Creating the foundation for organizational change (Monograph Series No. 5).* Action Learning Action Research Association (ALARA). https://www.researchgate.net/publication/259932785_Action_Research_Engagement_Creating_the_Foundations_for_Organizational_Change

Sarin, R., & Wieland, A. (2016). Risk aversion for decisions under uncertainty: Are there gender differences? *Journal of Behavioral and Experimental Economics, 60,* 1–8. https://doi.org/10.1016/j.socec.2015.10.007

Schein, E. H. (2010). *Organizational culture and leadership.* John Wiley & Sons.

Schwalbe, M., & Wolkomir, M. (2001). The masculine self as problem and resource in interview studies of men. *Men and Masculinities, 4,* 90–103. https://doi.org/10.1177/1097184X01004001005

Sommers, W., Loehman, R., & Hardy, C. (2014). Wildland fire emissions, carbon, and climate: Science overview and knowledge needs. *Forest Ecology and Management, 317,* 1–8.

Sontag, S. (1966). The anthropologist as hero. *Against interpretation and other essays* (pp. 69–81). Picador Press.

Stewart-Patterson, I. (2008). *Decision making in the mountain environment.* International Snow Science Workshop Proceedings. https://arc.lib.montana.edu/snow-science/objects/P__8107.pdf

Stoddart, M. (2011). Constructing masculinized sportscapes: Skiing, gender and nature in British Columbia, Canada. *International Review for the Sociology of Sport, 46*(1), 108–124. https://doi.org/10.1177/1012690210373541

Stringer, E. (2014). *Action research.* SAGE Publications.

Surowiecki, J. (2004). *The wisdom of crowds: Why the many are smarter than the few and how collective wisdom shapes business, economies, societies, and nations.* Doubleday.

Taylor, V. (1998). Feminist methodology in social movements research. *Qualitative Sociology*, *21*, 357–379. https://doi.org/10.1023/A:1023376225654

Thiollent, M. (2011). Action research and participatory research: An overview. *International Journal of Action Research*, *7*(2), 160–174.

United States Forest Service. (2016). *Sierra tree-strike incident: Learning review status report*. USDA, United States Forest Service. https://www.wildfirelessons.net/HigherLogic/System/DownloadDocumentFile.ashx?DocumentFileKey=f5012e97-557e-e64f-6e31-529ac2b34f93&forceDialog=0

Walker, E., & Latosuo, E. (2016). Gendered decision-making practices in Alaska's dynamic mountain environments? A study of professional mountain guides. *Journal of Outdoor Recreation and Tourism, Risk in Outdoor Recreation and Nature Based Tourism*, *13*, 18–22. https://doi.org/10.1016/j.jort.2015.11.010

Wieland, A. (2012). *Gender and decision-making: Competitive, risky and entrepreneurial decisions: Three essays related to how sex and gender influence decisions in different contexts* [Doctoral Thesis, University of California].

# Reflection on Chapter 6

## Where is the Climbing Ranger?

*Alison Criscitiello*

***Alison Criscitiello** is an ice core scientist and high-altitude mountaineer. Criscitiello's research explores the history of sea ice in polar regions using ice core chemistry. She drills ice cores in Antarctica, Alaska, Greenland, the Canadian high Arctic, and Yukon. She is Director of the Canadian Ice Core Lab (CICL) at University of Alberta, and an Adjunct Assistant Professor at the University of Calgary. Criscitiello holds the first PhD in Glaciology ever conferred by MIT. Criscitiello is a former US Climbing Ranger and has guided expeditions to major peaks in the Andes, Alaska, and the Himalaya. She has been the recipient of three American Alpine Club climbing awards, the Mugs Stump Award, and John Lauchlan Award. Alison has been named a National Geographic Explorer, and a Fellow of the Explorers Club and the Royal Canadian Geographical Society. Criscitiello is founder and co-director of Girls on Ice Canada.*

"Where's the Climbing Ranger?" was one of the most commonly asked questions of me during my younger years as a US Climbing Ranger in the National Park Service. It prepared me well for the similarly posed question years later, when I was a mountain guide—"Where's the guide?"—and equipped me to respond with a strong internal "yes" when, over and again, I questioned whether I was cut out for a PhD at MIT. Three similar experiences across three male-dominated disciplines conflated competence with masculinity.

Something long before my arrival at the places I've called home has shaped a clear pre-existing narrative around who belongs in those places. The discord between my whole-being sense of belonging and the simultaneous language of others denying that belonging could have been damaging. It could have changed the course of my life; I saw it change the course of women's lives around me. My reaction to this discord was to punch the bruise. Instead of running from uncomfortable situations, I made choices that potentially increased the number of those unnerving encounters. My second year in the National Park Service, I dyed my hair bleach-blonde and spiked it. I stopped shaving my legs. I chose to patrol

DOI: 10.4324/9781003089209-13

alone when at times I had the option of patrolling with another ranger. I pierced my lip and bent bike spokes into large hooped earrings that I wore on duty. I climbed consistently solo or in fast rope teams of two over complex glaciated terrain, both on patrol and in response to incidents.

During my years as a ranger, I had one person in my corner at work, and I have always thought that he made the difference between me pushing back against any resistance I met or backing away from the work I was doing. He was the Superintendent of my subdistrict in Olympic National Park, and he didn't see gender. He didn't see sexual orientation. He didn't see anything except his strong, independent, competent rangers and the work they did. He treated me as an equal, he let me make my own decisions, and he was always transparent and direct. He gave me an enormous amount of responsibility, which taught me from the start that he trusted me fully. It's hard to say what my experience during those years would have been like without him and what he represented. Though I'd like to think I would have persevered in that world regardless, I see there was huge power in having my larger-than-life male boss treat me—a 5'1" queer young woman— as an equal. During my second year as a Climbing Ranger, I made a solo rescue of another ranger, for which I received the Most Valuable Backcountry Ranger of the Year award. This was a pivotal moment for me. Though doing what we love is not for recognition, there is huge value—particularly within this context of not belonging—in being shown you are valued, your work is valued, and none of that value is contingent on irrelevant characteristics like gender or sexual orientation. From that moment on, I was more emboldened to rise. Not to just continue the type of work I had been doing, but to surge to the top.

I found that my work as an Apprentice Mountain Guide, on Denali and Aconcagua, was not dissimilar from my formative rangering experiences. Resounding is Reimer and Eriksen's finding that, "Men are presumed competent whereas women need to initially prove competence." I will never forget waiting at the Mendoza, Argentina airport to pick up three male clients who were flying in to climb Aconcagua. The first one to see me walked right up to me, the guiding company I was working for emblazoned across my shirt, and asked, "Where's the guide?" perhaps assuming I was simply there for airport pickup. I see in hindsight how I internalized this lesson, with its implication that a petite woman could not possibly be the mountain guide, though at the time the confluence of external microaggressions and undermining with internal headstrong determination was not so clear. I became very, very bold, and very determined to prove myself. I needed to be the fastest, the strongest, to carry the most weight. I needed to do things none of them could, in order to somehow show I was competent. I had to be capable of *more* to be equal. With those three particular clients, I proved this all to myself by summiting Aconcagua solo after bringing them back down to our high camp so they could rest after their failed summit bid. Importantly, this was made possible by the person who was the lead guide with me on the trip—the owner of the company—who saw huge value in me seeing the whole route, and who never questioned my ability to run up solo. I had a similar experience my very first time working on Denali.

When I left the Park Service and guiding, it was to go to graduate school to begin more deeply studying the places I loved to be. Frozen, high, harsh places. Places that I watched change over the years from the influence of warming surface temperatures. Glacier Meadows, at the toe of Blue Glacier on Mount Olympus, was my changing, shifting heart. Increase in avalanche and rockfall across the Olympic Mountains during my time there was obvious and impactful to me. While grappling with sense of place and belonging, I was also witnessing mass loss on glaciers that I knew well. It was something I wanted to understand better, and to study. Even this early on, I was watching the seeds of climate change's gendered impacts germinate. The increase in avalanche- and rockfall-related incidents I was responding to were situated within a larger context of heightened risk in travel across these mountain landscapes due to warming surface temperatures. If risk assessment, as well our transparency in assessing risk, is gendered, would mountain travel not also become even more polarized by gender? A willingness to acknowledge risk and be transparent about its assessment becomes increasingly important as our natural environment shifts to an inherently riskier one.

Perhaps owing in part to the crucial people peppered throughout my younger years, such as my boss in Olympic National Park, I don't think I was ever at risk of the "belief that [I was] to blame for [my] own experiences of discrimination," as described by Reimer and Eriksen. These crucial mentors along my path were my bumpers, keeping me from derailing in either direction—by either abandoning the work I was doing due to the external messaging I constantly received, or doing the work but in a fatally risky way. My personality lends itself toward the latter, but the train never left the tracks. Mentorship may be one tool to help change the gendered landscape of both mountain and hard science environments.

Fostering such mentorship in an organized way and growing something from seed has been the only way I have found to address, in my own work, the root of the "gender-environment nexus." In 2018 I co-founded Girls on Ice Canada with three other Canadian women. Rooted in cultivating self-confidence, leadership, and curiosity, its founding felt like a defiant act. Reimer and Eriksen ask, "What does embodied feminist leadership in the mountain and wildland environment look like? Can we imagine—or perhaps reclaim—these environments as a non-gendered space that osters the emergence of tolerant and inclusive professional mountain and wildland cultures?" Here, we've tried to do just this by offering a cost- and barrier-free science and wilderness immersion programme for female-identifying high schoolers led by female-identifying scientists, certified mountain guides, and artists. One of our organization's core values is centred around the idea of belonging. We believe that individual differences strengthen our community, and this includes embracing the evolving definition of girls. We nurture relationships based on inclusion, diversity, and equity.

Risk tolerance has an interesting place within all these spaces. Through instructing Girls on Ice Canada expeditions, I have noticed (among many differences between all-female and mixed-gender teams in the mountains) that within all-female teams, there is no need to hide risk-taking or the process of assessing risk. I have to think this is partly why it has been my choice to climb only with women for all personal big expeditions to date. A team of all women does not have to navigate the exhausting effort of concealing risk or the implications of risks taken when there's no other option available. Especially as a new mom, I have experienced additional criticisms for perceived risk that I take in the mountains and within my job. Reimer and Eriksen note how "workers who were also mothers […] described being treated as if motherhood compromised their risk management abilities and competence." Like many, my assessment of risk—of what is appropriate to do with my own body—was suddenly questioned constantly upon giving birth. I said this to the first person to directly tell me they didn't agree with a mom being in big mountain terrain, and I will continue to say it: To me, the biggest risk as a mom is to stop being who I am in front of my daughter. The most critical thing I can do is keep being me. The last thing I would want to teach my daughter is that bringing her into our lives has been mutually exclusive with being my whole self. Of all people in the world, I want her to know who I am.

The blazing truth is that I still do all of this. I still pressure myself to be the fastest, the strongest, I need to carry big loads and show that I belong. I suppose the difference, after all this time, is that I'm proving (to myself) that I belong in the lead. I put similar pressures on myself with scientific funding, with leading as many large projects as I can at any given time. I am the Director of Canada's national ice core lab, yet I am still regularly gaslit. In insidious, underhanded ways, I'm manipulated into questioning my own expertise and position at the top of my field. Recently, when I was cast as a character in a film that will be showing within an ice core exhibit at my city's science centre, the first draft of the male-written script had me sounding like I was anything but an expert. I read the draft script with a dumbed-down voice that was supposed to be my own and felt a sinking in my gut. It took a couple reads to realize not only did this not sound like me, it didn't sound like any expert in any field. The onus, of course, was on me to address the obvious truth that sexism drove the voice of this character who was not, in fact, anything like me. Even with a feeling of nothing to lose, I find I am still driven by these lessons I have learned over a lifetime of orbiting. Of feeling confident, competent, and at home, but circling around the belonging at the heart of it all.

# 7 Contemporary Feminist Analysis of Australian Farm Women in the Context of Climate Changes

*Margaret Alston, Josephine Clarke, and Kerri Whittenbury*

**Margaret Alston** *is Professor of Social Work at the University of Newcastle and Professor Emeritus at Monash University, Australia. While at Monash University, she established the Gender, Leadership, and Social Sustainability (GLASS) research unit. Her research and teaching are focused on gender, climate and environmental disasters, rural women, and social work, including research projects with Oxfam and the Australian Centre for International Agricultural Research (ACIAR). She has been engaged as a gender expert with the United Nations Environment Programme (UNEP) and the Food and Agriculture Organization (FAO). In 2010, she was awarded an Order of Australia Medal (OAM) for her services to social work and to rural women. In 2021, she was made a Member of the Order of Australia (AM) in recognition of her work in social work and with women. Professor Alston is a past-Chair of the Australian Heads of Schools of Social Work (ACHSSW) and was appointed a Foundation Fellow of the Australian College of Social Work in 2011. She is currently completing several gender-based projects.*

**Josephine Clarke** *is Research Associate in the Social Work Innovation Research Living Space (SWIRLS) at Flinders University. Dr. Clarke has a background working and researching in rural and remote communities and has completed research on a range of issues including rural gender relations, livelihood and industry restructuring, climate change, and water reform.*

**Kerri Whittenbury** *is Adjunct Senior Research Fellow in the Department of Social Work at Monash University, Australia. Dr. Whittenbury is a sociologist with research interests in women and gender, rural communities, rural and farm families, irrigation communities, and gender and climate change. She is currently working on several research projects that focus on the gendered social and health impacts of climate change and drought.*

DOI: 10.4324/9781003089209-14

*Note: This article has been adapted from its original version: Alston, M., Clarke, J., & Whittenbury, K. (2018). Contemporary feminist analysis of Australian farm women in the context of climate changes. *Social Sciences*, 7(2), 16. https://doi.org /10.3390/socsci7020016

## Introduction

Climate changes are reshaping agricultural production and food security across the globe (Intergovernmental Panel on Climate Change [IPCC], 2014b), leading to increased uncertainty (Torquebiau et al., 2015), changing production processes (Bryant et al., 2016), greater levels of outmigration from rural areas for alternative income (Alston, 2015), and a reshaped agricultural workforce (Preibisch & Grez, 2010). As these impacts take hold, significant gendered workforce realignments are underway in agricultural production units. One result evident across the world is that women are having a much greater role in food production and now comprise 43% of the global agricultural workforce (Food and Agriculture Organization [FAO], 2013; World Bank, 2012; World Bank, 2017), over 50% in many Asian nations, and over 40% in southern Africa (FAO, 2010; World Bank, 2017). Nonetheless, there is far less documented information on women's contribution to agriculture in developed countries and almost no systematic data collection in Australia. This is despite the fact that 99% of the 134,000 Australian farms are run as family farms (National Farmers' Federation, 2012), confirming that Australian agriculture is highly dependent on the labour flexibility female and male family members provide.

Yet, agricultural industries in the developed world operate in a policy, media, and industry environment that is hegemonically masculine (Knuttila, 2016) and that rarely acknowledges the significance of women's input to successful farm production. Consequently, although women's work is acknowledged as crucial in the private family sphere (Alston & Whittenbury, 2014), women have struggled for public recognition of their contributions and may feel powerless to shape their own destiny. Consequently, when radical changes such as climate events occur, forcing significant adjustments to agricultural production processes and major gender renegotiations around workloads, women and men enter these negotiations from vastly different power positions (Shortall, 2006). Thus, while documenting agricultural work by gender in developed countries is long overdue, documenting the adjustments resulting from climate-induced environmental events would seem critical.

To examine how women's labour on farms is being reshaped in the context of both climate changes and a historical invisibility, this chapter focuses on one example of an environmental crisis that is reshaping gender relations on Australian farms. The Murray-Darling Basin (MDB) area of Australia is a large, traditionally highly productive area, known colloquially as the nation's food bowl. This area is experiencing significant problems associated with water availability for irrigation farming following long years of drought and a historical over-allocation of water licenses. This has resulted in major ecological and river health issues, leading to government policies being introduced to recover water for the environment, which in turn reduces the amount of water available for irrigation farms in the Basin. Through a series of measures, water savings have necessitated sometimes radical changes in farm operations and this has had significant impacts on irrigation family farms across the Basin (Alston et al., 2018). In a previous paper (Alston et al., 2017), we noted that these negotiations have resulted in the emergence of a "farmer-manager" role occupied by male farmers while women appear to be taking on a more "directed worker" role. We noted that women's work remains far less visible despite the fact that they appear to be increasing their workloads both on farms and away from the farm, securing much needed income through their off-farm work.

In this chapter, we further explore gendered labour adjustments by documenting gendered changes in workloads on Australian family farms subsequent to the introduction of environmental water saving policies in the MDB. Given the historical invisibility of women in agriculture in developed countries, we provide a particular focus on women's views in the context of their farm and community. We argue that despite women's significant input to agricultural production on these farms, many continue to view themselves as agricultural outsiders. Before turning to our findings, and to contextualize women's agricultural labour, we discuss the rich and ongoing feminist analysis of farm women in developed countries and outline global climate change experiences and predictions.

### Women in Agriculture: Developed World Analysis

For over three decades, feminist analyses of farm women have addressed the invisibility of women engaged in agricultural production in developed countries (for an early feminist analysis see for example Rosenfeld, 1986). As we wrote in 2010 referring to feminist research on agriculture, these earlier researchers "addressed a set of issues: the gendered division of labour within households, women's identities as farmers or farm wives, the economic contribution of women through their on- and off-farm work, women's access to land and capital, and the virtual invisibility of women in agriculture" (Sachs & Alston, 2010, p. 277), issues that continue to be relevant. Critically, feminist theorists noted that family farms had previously been viewed as largely undifferentiated production systems in which women were "invisible farmers" (Sachs, 1983; Williams, 1992). Thus, they argue that there has been very little acknowledgement of women's work other than as a largely hidden "factor of production" (Whatmore, 1991, p. 5). Farm women's identities continued to be constructed around their most common entry point to agriculture: Through marriage, and the notion of "wifehood" has tended to shape the expectations of their role, particularly in relation to caring for, and enhancing, the wellbeing of other family members (Sachs & Alston, 2010, p. 279).

More contemporary feminist analyses of farm women across the developed world continue to expose the critical nature of women's work to a farm enterprise, by focusing on the significance of gender to farming and the deeply embedded gender relations that shape, and are essential to, family farming. In so doing, feminists have facilitated a healthy critique of farm women's invisibility (Riley, 2009); of the ideologies that prioritize male power and influence (Brandth, 2002a; Price & Evans, 2006); of hegemonic masculinity (Knuttila, 2016); of the patri-archal power and male privilege that shape both agriculture (Shortall, 1999) and gender identities on farms (Brandth & Haugen, 1998; Haugen & Brandth, 2017); of power, agency, and resistance (O'Hara, 1998; Shortall, 2017); of the masculi-nation process and its links to technological developments (Brandth, 2002b); and of rural customs, such as patrilineal inheritance practices and farm organizational leadership roles, that continue to disempower women (Alston, 2000; Bock & Shortall, 2017).

While this research suggests that women are having a much more visible role in global food production, there are a number of factors that reduce the acknowl-edgement of their labour. Chief amongst them is the issue of land ownership, which is one of the key facilitators of the invisibility of women in farming. Patrilineal inheritance practices have dominated global agricultural resource own-ership, leaving women's chief point of entry to farming to be marriage. Globally, women own less than 10% of land, receive only 5% of all agricultural exten-sion services, and have access to only 10% of total agricultural aid (FAO, 2013). In Australia, gender-disaggregated data on land ownership is scarce; however, research suggests that women inherit farmland in only 5–10% of cases (Australian Broadcasting Commission [ABC] Rural, 2016; Dempsey, 1992). Yet, as Shortall (1999) notes, "property is power" and, because women rarely figure in agricultural land ownership, they have far less power in farm family agricultural units. At its most extreme, this lack of ownership and support can translate to reduced access to household goods and to greater food insecurity for women and girls (Alston & Akhter, 2016; International Fund for Agricultural Development [IFAD], 2015; Pinstrup-Andersen, 2009). At the very least, it leads women to feel keenly the lack of acknowledgement of their work.

Research from several developed countries reveals that women's labour remains a taken-for-granted, largely invisible factor in a structural system where gender relations are dominated by patriarchy and male privilege, and women's work remains essential but masked (Fletcher & Kubik, 2016). Sheridan and Haslam-McKenzie (2009, p. 9) note that women's invisibility in plain sight is partly explained by their role straddling the private world of the farm and the public world of the community: The "space of betweenness" that reduces their legitimacy as farmers.

Additionally, could we not also argue that men are equally trapped in the pre-scribed role of the "good farmer" (Haugen & Brandth, 2017), a role that when not fulfilled—for example when drought erodes agricultural expectations as it did in the Basin—it can lead to mental health consequences (Alston, 2012)? Does their "power" have a downside when times are so tough? Perhaps we might also ask could climate change, when linked to climate-related policy developments and increases in women's agricultural labour, be a point of departure from women's invisibility? Will these changes give women increased agency and relieve the pressures on men to be exemplary, stoic, "good" farmers? Or will recognition of their efforts be dampened by the lack of gender mainstreaming of policy and a failure to attend to the gender implications of policy developments and national responses to climate change in developed world economies? As Prugl (2010) notes, the state is Janus-faced and has the capacity to reproduce or challenge gender inequalities. It could potentially do this through gender mainstreaming of policies to assess the types of supports required by women and men to make the significant changes required by climate changes.

However, with the Murray-Darling Basin water reform process, the state has not only stood back from actions that might assist equality, it has also been ruthlessly gender-blind, and has committed to policies that appear to run counter to gender equality. Largely, this is because the neoliberalism underpinning policy developments in Australia have led to a retreat by government from any form of subsidization of agriculture; a hands-off approach to policy development (Alston et al., 2016); and reliance on markets to achieve environmental and social outcomes (Hussey, 2014). These processes in turn place pressure on individual farm units, and it is left to them—and the people who work there—to adjust to market forces by shaping their enterprise and labour accordingly. How then do the neoliberal market-based principles that underpin the Murray-Darling Basin Plan affect gender relations and labour allocation on Australian farms? Before turning to a discussion of the research, we describe the area that formed the basis of the study and examine climate change predictions.

### Murray-Darling Basin Area of Australia

The Murray-Darling Basin is a large, highly productive area that extends across much of the eastern area of Australia. It covers over 1 million square kilometres, or 14% of the country, and 23 river valleys and produces 40% of Australia's agricultural produce (Murray-Darling Basin Authority [MDBA], 2010). Over 2 million live in the Basin and a further 1.3 million live outside the Basin but are dependent on Basin water (MDBA, 2010). In the Basin, there is one major city, nineteen urban centres with populations greater than 10,000, 159 small towns, and 230 small rural locations. A majority of these communities are reliant on agriculture (MDBA, 2010). In addition, a significant proportion of the population, particularly in the smaller communities, are employed in agriculture or food processing. The Basin sustains a variety of industries, many dependent on irrigation water. Thus, when environmental scientists expressed concerns about river health and ecosystem damage in the Basin following years of drought, this raised significant concerns across the Basin.

Previously, the state governments had been responsible for water; however, the problems identified in the Basin forced the national government's hand and the 2007 *Water Act* (and its 2008 amendment) gave water responsibility to the federal government. Consequently, a number of water policies have been introduced by the federal government to increase the amount of water available to the environment. This necessarily requires the retrieval of water from agriculture, and hence there have been major disruptions in the Basin. However, it is important to note that despite a lack of trust in governance processes and increased uncertainty amongst water users, there is general acceptance of the need to address water concerns amongst all stakeholders, including irrigation farmers (Alston et al., 2015). The Murray-Darling Basin Plan (MDBA, 2012) was released in 2012 after much disquiet being expressed by Basin people. Irrigation farm families and their communities were concerned about the fairness and justice of the processes to be implemented to save water.

Subsequent water policies have led to schemes, such as the buy-back of water licenses from farmers and the introduction of efficiency measures, to improve irrigation infrastructure and reduce water wastage. These have required significant investment by government, major adjustments by industries, and changes in farm production units across the Basin. In 2014–2017, we undertook a major study with dairy farm families in the southern parts of the Basin to ascertain how these changes were impacting the social sustainability of Basin communities. Arguably, this case study reveals the way climate-related drought conditions and a consequent need to deliver water to the environment have had a major impact on food production processes. This research provides a useful example of adjustments made in response to climate change. In this chapter we narrow our focus to the impacts of a climate-related event, and of water markets on gender changes in workloads on farms.

### Climate Change in the Basin

There is no doubt that climate changes are having a significant impact on food production across the world and this will continue to affect farm production units. The Intergovernmental Panel on Climate Change (IPCC, 2014a, b) predicts that the earth will become hotter and drier, that there will be more extreme weather events, increased temperatures, and a rise in sea levels across the globe. More frequent heat waves and extreme rainfall events will affect many parts of the world, and these will extend for longer periods, resulting in a negative impact on crop yields in many regions (IPCC, 2014a). Without attention to emissions reductions, the IPCC predicts that there is a significant likelihood of more severe and irreversible outcomes (IPCC, 2014a). The impact of these changes on food security will be critical. The IPCC (2014b) notes that there is a high risk of disruption to water supply and food production and that this will cause food insecurity, a breakdown in food production systems, and disruptions to food access, food utilization, and food price stability across the world.

Climate-related research in the MDB supports these global contentions. Researchers suggest that the temperatures in the Basin have increased by up to 1 degree since 1910 and that this is predicted to rise by up to 1.5 by 2030 (Timbal et al., 2015; Whetton, 2017). This will lead to hotter days, increased periods of drought, less frosts, soil moisture and rain, and a harsher bushfire environment.

It should not be surprising then that climate-related events are changing agricultural industries, practices, and the agricultural workforce in the Basin and are accelerating socio-economic trends, such as a reduction in full-time employment (To, 2017), the outmigration of rural workers seeking alternative incomes for their families, together with a greater role for women in farm labour. In this chapter, we note that while gender power relations persist, climate changes are reshaping work practices in agricultural units in developed countries along gendered lines. Our research with women and men on irrigated dairy family farms in the Murray-Darling Basin area of Australia provides one example of these changes.

## The Research

Our research was conducted in the MDB area from 2014 to 2017 with irrigation dairy families and communities located in the south of the Basin. The project was funded by the Australian Research Council and the partner organization, the Geoffrey Gardiner Foundation. The purpose of the project was to examine factors shaping ongoing social sustainability in the Basin following the impacts of the Basin Plan. Three communities were chosen for detailed study and these communities had been defined as especially vulnerable to the proposed reforms in the draft Plan (MDBA, 2010). The project proceeded in four stages and included 90 semi-structured interviews and focus groups conducted during 2014 to 2016. The first stage included key informant interviews with regional, state, and national informants who were critical of the reform process, and these were designed to gain an understanding of the policy process and its implications and goals. The second stage involved 25 interviews with local key informants who had expertise relating to their communities. These were designed to gain a more informed understanding of the impacts at community levels. The third stage involved 43 interviews and focus groups with dairy family members who were asked about the impacts at the farm and family levels. The final stage involved a partnership arrangement with the University of Canberra's Regional Wellbeing Survey to incorporate surveys of dairy farmers in the areas that formed the basis of our research. In total, 128 surveys were returned by dairy farm family members (31 women, 93 men, 4 missing values). Qualitative data were thematically analyzed using NVivo and this allowed the research team to draw out dominant themes across all interviews and to assess the factors that are assisting or reducing adaptation to the water reform process. Quantitative data were analyzed using SPSS, facilitating the development of frequency tables and further statistical analysis. This paper draws on stages 2, 3, and 4 and focuses only on gender implications of the climate-related water policy interventions at the farm level.

## Findings

A number of issues emerge from this research that suggest that climate changes are creating significant realignments in gender relations on farms. As we have noted previously (Alston et al., 2017), the emergence of the "farmer-manager" role undertaken by male farmers and the "directed worker" role undertaken largely by women is evident. What is also apparent, and the argument that this chapter develops, is that there appears to be a significant increase in women's work hours on- and off-farm and this affects the way women view themselves and their roles on farms and in their communities. Yet, because this receives little recognition beyond the farm, there is limited acknowledgement of these changes and little policy development to address the issues this raises for women. Further, we note that women and men going through significant adaptations to climate-induced changes report significant health issues and these provoke further changes in gender dynamics.

### *Changing Work Roles*

Our research reveals that there have been major changes in women's and men's work on farms and Figure 7.1, outlining results from our quantitative survey, gives some indication of the way these changes affect both women and men. The change appears to be greater for women, many of whom have increased, or intend to increase, both their on- and off-farm work. More than half of the female survey respondents reported they had, or intended to, increase their off-farm work and their on-farm work. At the same time, a high proportion of women and men reported that the amount of hired labour on farms had been reduced, suggesting that more work is being absorbed within the family. Adding to the complexity of job-seeking off-farm is the fact that both women and men felt there were few jobs available locally and this will have a critical impact on access to paid work off the farm. Figure 7.1 illustrates these issues.

As noted previously (Alston et al., 2017), our qualitative interviews give additional insights into how individual women's lives are being shaped by their increased workload. The following is from a young dairy farm woman who reported her workload as follows.

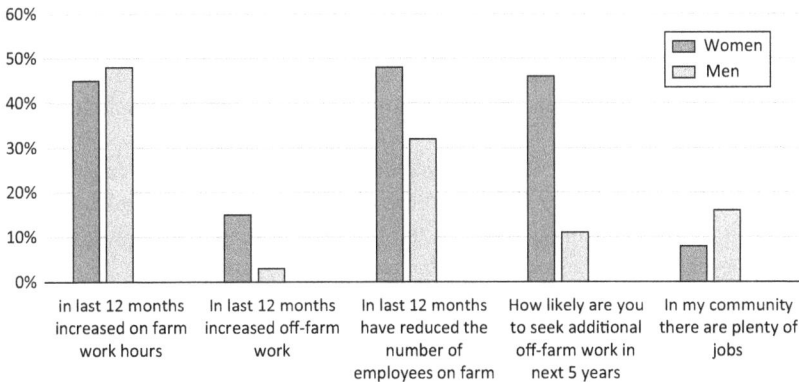

*Figure 7.1* On- and off-farm work: Reality and perceptions.

> I do the most milking, [partner] helps me and we also have two others, we have [partner]'s mum that milks as well so I work with her predominantly and then we have people on the weekends to give her a spell... I do whatever else he tells me to do, which might be drenching cows, calves, seeding... So, it's a lot, the bookwork is insane, it's crazy... This is an 800 acre farm and we're milking 550 cows, and just milking cows alone is 35 h, so 35 h of my week I'm just milking cows before I do anything else, before I do any book-work, before I do anything on the farm, and I'm still doing stuff on the farm, so, you know, it's just a lot, and the kids, so just everything. (Stage 3 C1 10).

Other women reported on their off-farm work contributions, many noting its significance to the day-to-day survival of the family.

> I'm working off farm because it does put food on the table, and that is a good thing. That is a good thing... But as far as running the farm's concerned it doesn't touch the edges. Not even close, it wouldn't even touch a quarter of the interest. It might keep food on the table. And, it keeps me sane, because as much as it's interesting to talk about the water I couldn't do it all day every day. (Stage 3 C1 5).

Older women sometimes noted the transition from hands-on work to running the business management side of the farm as a stage of life transition.

> Well, originally when we first started the farm I milked seven days a week and then did the farm work in the middle of the day and then did all the book-work at night, and now I only fill gaps on the farm in terms of the physical roles, ... Now I just do the financials, so that's so that's my main role and to keep on top of everything it really takes about three days a week to keep up with everything and the wages and the superannuation and any other staff issues, and paying bills and doing cash flows and keeping your finger on the pulse of where all that's going and dealing with the banks and those sorts of things. (Stage 3 C2 2).

Several women noted that farm work tasks are re-defined as domestic labour when taken over by women and therefore as an extension of their household tasks.

> So, I think it is hard for women to see themselves as an equal because we're seen as the wife, raising kids, washing clothes, and calf rearing, that's the normal traditional woman's job on the farm.... Oh yeah, and hosing out the [dairy] yard, that's the other girl job. (Stage 3, C1 5).

Men also refer to their increased workloads.

> I'm out half the night watering... so that's your tipping point, and you can't put labour on to water at night. Because you already know where the water's

going from the day time you tend to be the one who goes around it again at night, and you can't get a labourer to do it at two in the morning to go around and change that bay from one to the other, you can't. (Stage 3, C1 07).

### Economic Pressures

The changes in the Basin, and in the dairy industry more generally, have had an impact on dairy prices and on the capacity of individual farm families to reshape a viable production unit. Farm families continue to adjust to the new realities, yet key informants confirm the economic pressures on families. This has had a significant impact on the way farm family members feel about their work and this too has a gendered dimension.

As illustrated in Figure 7.2, women dairy farmers were more likely to report feeling less positive about the viability of their communities and hence the liveability of their environments. This translates into negative views held by women and men on the capacity of the Basin Plan "architects" to understand their experiences. Women are less likely to feel positive about the community and local economy but were more trusting of government understanding their experiences.

### View of Themselves

In both the survey and interviews, women were more likely than men to report feeling like an outsider (18% women, 4% men), and were less likely to report feeling welcome in their community or to be part of their community. Figure 7.3 illustrates these points. This suggests that, because of their entry point to farming, and the consequent uneven power relations, women struggle with identifying themselves as farmers despite their workloads.

Qualitative data support this view, with women commenting on the "blokeyness" of both farming and farming communities.

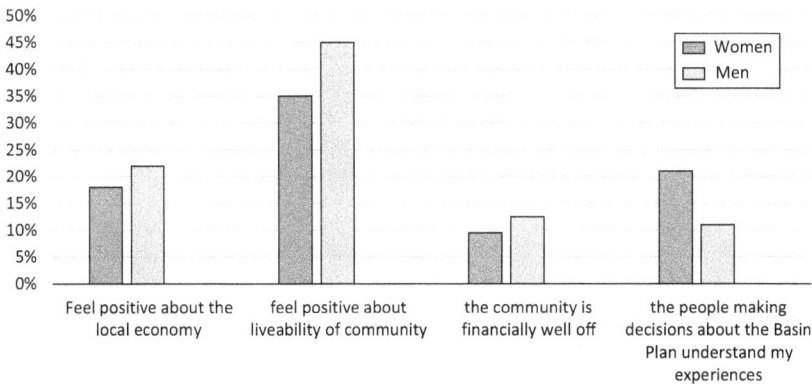

*Figure 7.2* Views on the local economy.

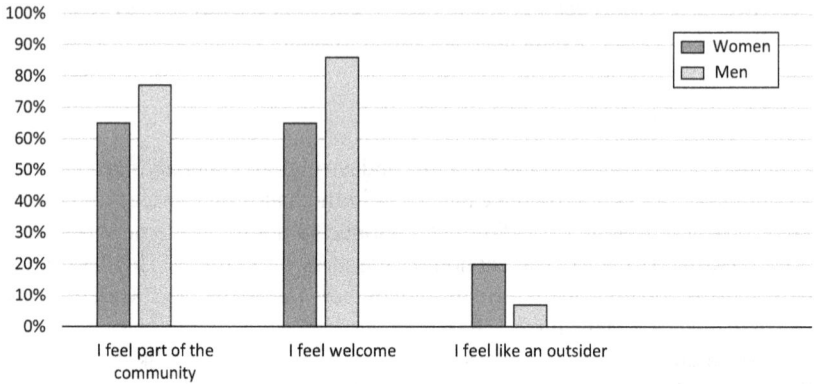

*Figure 7.3* Views on the community.

> Most of the women that I know have got young children so they're sort of balancing feeding calves and looking after kids with not much support from the men. They are quite 'blokey' around here... it is really, really sexist. (Stage 3 C1 5).

Others refer to the difficulties they have defining themselves.

> I used to introduce myself, 'I'm a dairy farmer's wife,' and so [I say to myself], 'No, you're a full time professional dairy farmer, that's what you should say you are, and when you meet men you shake their hand and say hello'. I shouldn't have to do all that stuff really, I should be treated as equal but it is hard in probably a male dominated industry to step up, but there are some pretty good women who I talk to. (Stage 2 C1 KI 7).

As illustrated in Figure 7.4, while there is little difference in the expectations of women and men about being able to achieve life goals, there are marked gender difference expectations relating to work, optimism about the future, and feelings about the farm. The difficulties experienced by women in trying to define themselves appear to translate into lower expectations for women, lower levels of optimism, less positive views of the farm's future, and less certainty that they will be able to achieve what they had hoped in their life and in their work.

### Health Consequences

One of the most evident issues for farm family members going through the uncertainty of climate changes, the policies developed to address them, and drastic changes in livelihoods and labour is health outcomes. Because women appear to be more conscious of the health of family members, this issue has a significant

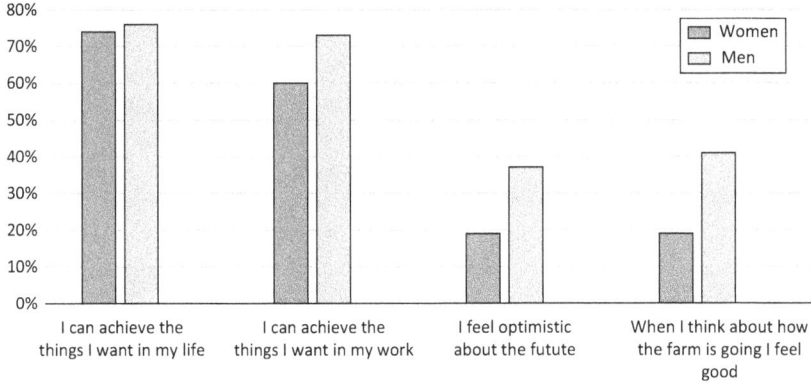

*Figure 7.4* Achievement and optimism.

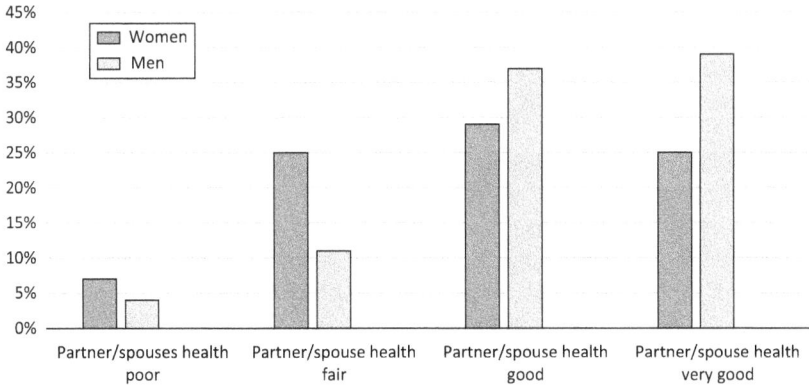

*Figure 7.5* Rating partner's health.

impact on gender relations and the compliance of women when family members are suffering. Women are more likely to discuss their men's health and ignore their own, and they tend to rate their men's health more poorly than men rate women's health (see Figure 7.5). Men rarely discuss women's health in interviews and are more likely to report women's health status more positively. This suggests not only that women continue to ignore their own health issues, but also that there are gendered differences relating to health and this adds to the complexity associated with livelihood adaptation.

Service providers within the communities also commented on the health of families.

> We certainly found in the drought, the men didn't often do much talking but the wives did, because they're the ones that often carry the burdens—they

carry it more on their sleeve and the guys tend to just go in the paddock and do what they've got to do. (Stage 2 C3 KI 3).

Well, the impacts were obviously isolation, and depression was one of the big ones that we saw, and particularly back in the early to mid-2000s thereabouts, or... from probably 2001 through to about 2008 there were quite a few suicides across the region, we were seeing more and more wives presenting for health reasons, because the men didn't present because they were too busy trying to either save what they had or buried themselves in their work. So, I guess overall health generally was what presented and obviously mental health was one of the big key themes. (Stage 2 C3 KI 3).

One service provider pointed out:

We had a young fellow ring us up, who was just at his wits end, he'd been everywhere, they'd taken his water off him, and that's according to him, they'd taken his water off him, his wife and kids had left him. (Stage 2 C2 K1 2).

Health workers, education providers, and other key informants also commented on the impact on relationships, with many noting a rise in relationship breakdown, family violence, mental health issues, and the use of drugs and alcohol in their communities. In the survey, women were more likely to report that there is alcohol abuse (64% of women as opposed to 40% of men) and drug abuse in their community (56% of women as opposed to 39% of men).

What is evident from the interviews with both women and men is that men are experiencing significant mental anguish at the same time as women are actively monitoring the health of their partner. This supports the idea that men strive to be "good" farmers and when they cannot achieve the ideal standard their mental health deteriorates. Women referred to how extreme this behaviour could be. Men withdraw, relationships suffer, and there may be self-medication with drugs and alcohol. Because women monitor their men closely, they may step in and take over farm tasks for periods of time as the following quotes indicate.

My husband was extremely depressed probably 12 months after we got married because it was really, really hard and financially it was very tough, and so I took over the managing of the farm and he would just milk the cows and drive tractors. (Stage 2 C1KI7).

My husband he may as well have died, I reckon, nearly. He'd stopped milking and walked away, and didn't pack anything up, it just looked like that. Now he would hate me saying all this, by the way, so lucky his name is not the same as mine. So yeah, that's what it was like, it was devastating, and then it took about probably three years before he could even go back there to look at it because every time he went he would get so angry and devastated and watching it all fall apart, he just couldn't handle it. (Stage 3 C1 10).

As a result, it was not surprising that women were more likely to note the very real issues associated with a lack of specialist services, including mental health services.

> Out here, between [town name] and [town name], no one can hear you, no one can see you properly, you've got to drive a couple of k's down the drive-way, and I think that's really scary. So, there's a lot of hidden stuff going on, and then the other professionals are almost scared and not knowing what to do. (Stage 3 C1 5).

When men discussed their health issues in interviews, they invariably referred to mental health factors.

> It's always that male thing, we always struggle with going to the doctor and stuff like that, I reckon. I know in this area especially there is a lot more sui-cide and stuff like that, and I've had people that I know that have been in that position, so that's a big issue, I think, mental health in general. (Stage 3 C1 4).

When women discussed their health, it was more likely to note the bodily ail-ments that accompanied the increase in physical labour, particularly those affect-ing knees, hands, and backs.

> We're milking 550, it's back breaking. Physically if I think about it, I'm milking those cows, it's three and a half thousand, four thousand times I'm doing this a week, that on your body, doing that, you get sore shoulders, back problems, it's hard, it's hard work. It's hard work but it's not, do you know what I mean, it just takes its toll on you. (Stage 3 C18).

As a result of the pressures on families, service providers were also likely to note that women are more likely to leave when the pressures are too great. This may be a result of their status as outsiders and "non-beneficiaries" of farm inheritance practices.

> The family breakdown that can happen because of the economic pressures, and wives will be more willing usually to cut losses and walk away than men will from farms... But on the whole it's the women that will say, 'I can't take this anymore, I just don't want to do it anymore', and its often because they're see-ing the impact on their children... so mothers are more likely to be the ones who will say, 'This is just not acceptable', and want out from farms. And, you can see immediately the problem when one wants out and one wants in you've got a breakup of a marriage and then you've got the breakup of the farm, and then you've got all sorts of huge issues there. (Stage 2 C1 KI 4).

In summary, while farm labour is reconfigured in response to climate events and subsequent policy changes, this is taking place against a background of complex health impacts. Women appear to be constantly monitoring family health, often while ignoring their own, and the changes have particular impacts on men's health and family dynamics. These complexities may make it difficult for women to address gender negotiations as they may be reluctant to upset family dynamics further by challenging the status quo.

### Government Response

The response of government to the climate-related environmental crisis in the Basin was to put in place complex water recovery strategies, and these addressed the identified crisis: The lack of environmental water. Governments acted in both an environmentally and economically responsible fashion and these actions were accepted by stakeholders as necessary and desirable. Yet what was not addressed were the social outcomes of these significant changes.

Thus, to the outside observer, there appeared to be no government departments that acted as "champions" for rural people and communities going through critical adaptations to climate-related stress. Further, there appeared to be no social analysis on factors such as employment, welfare, services, health, and other factors that shaped people's responses. There were no obvious community development processes, no fostering of alternative industries, and limited attention to telecommunications and transport infrastructure. There was no gender mainstreaming of water policies, no gender analysis of outcomes to determine the issues affecting women and men, particularly on family farms, and hence no focus on child care services, mental health provisions, and employment development and training programmes. Because there was no attempt to apply gender mainstreaming processes to water policies, there is limited institutional awareness of the gendered implications of the changes nor any analysis of the complex gender negotiations taking place on affected family farms.

As one industry leader noted:

> I still believe there was this massive denial that taking the water out wouldn't cause massive adjustment within communities, and I think the Commonwealth didn't want to recognize it and it is still happening today. They just don't want to face up to it as a consequence of their decision, they'll say, 'Oh well, the market will look after it, the water will trade from the low value users, so it won't have much impact', but it definitely had. (Stage 1KI 46).

## Discussion

While women's work on family farms in developed countries has been largely invisible, we argue that climate changes, and consequent workload adjustments taking place on family farms, continue to marginalize women. Climate-related changes, such as temperature rises, less rainfall, and more periods of drought, leading to the reduction in the amount of water available for irrigation in the MDB, have created the need for farm production units to restructure their enterprises. One result is the increased workloads on- and off-farm being undertaken by women. There has been limited attention to, or acknowledgement of, changing gender workloads in the developed world and minimal focus on these gendered changes in the context of climate-related events.

It is perhaps no surprise then that when women are drawn into agricultural production as a result of the complex interaction of climate changes, recast agricultural production techniques, and renegotiated gender work roles, women's input is treated as a "farm survival strategy" rather than a major personal economic contribution by women to the enterprise (Alston & Whittenbury, 2014). Their tendency to relieve their partners of various work tasks in order to lessen emotional distress is largely hidden. Women's increased involvement in agricultural labour is not necessarily a source of empowerment for women and appears to reflect the "feminisation of agrarian distress" that Pattnaik et al. (2017) identified in their research. Further, as women move into increased farm work roles, society "reinterprets women's new work through a heterosexual matrix" and gender relations are reshaped and redefined in hierarchical ways (Sachs & Alston, 2010, p. 279). Thus, we find various tasks redefined as an extension of household tasks or of women's nurturing role and not necessarily "work." In this way, as Prugl (2010) writing about German farm women notes, patriarchy constantly reproduces itself in new ways to fit new circumstances and altered configurations of unequal power. Ultimately, this has repercussions for women's capacity to influence not only the discursive constructions of family farming, but also power relations, labour allocations, and policies and practices relating to agriculture, climate changes, and rural communities that shape agricultural production on their farms.

Our research mirrors that of other male-dominated areas where women's work is undervalued. Drawing on our research example in the MDB, we have established that climate change-related events and agricultural restructuring have significant gender implications for women and men on family farms in the developed world. For women, it continues a historical ignorance of women's farm work, resulting in a lack of understanding, limited official acknowledgement, and a subsequent discounting of the labour contribution of women to agriculture. It also facilitates and continues the historical absence of women from leadership roles and fails to address the implications for service delivery and support strategies needed for viable adaptation. For men, the lack of services, in particular mental health services, can lead to increased social isolation and despair. For women, the lack of child care services and limited training opportunities limit their adaptive capacity. We therefore argue that the lack of gender mainstreaming in climate change and water policies has significant implications for women's continued invisibility and for women's and men's health and wellbeing.

We would argue the need for significant changes in the way women's contributions to food security in developed countries are assessed, valued, recorded, and supported. This will require governments to respond more effectively to all members of farm families in the context of climate changes. What is needed in affected areas, such as the MDB, is a whole of region development plan incorporating gender equality, community development, and a supportive service infrastructure to assist families adapting to changes that are beyond their control. It is insufficient to expect markets to be the sole determinant of people's capacity to adapt. As we have argued elsewhere (Shortall & Alston, 2016), this necessitates a "rural champion": A dedicated government department that focuses squarely on rural development, and one that incorporates attention to gender mainstreaming during increasingly prevalent climate-related changes.

When we view this research in the context of farm women's historical invisibility, it is perhaps not surprising to see complex gender negotiations occurring in the private family sphere being ignored by industry and government. The history of farm women is peppered with extensive research indicating not only their invisibility, but also the continuation of the public view of farming as a male occupation. Thus, women's efforts go largely unnoticed beyond the farm gate and this creates significant personal tensions for women. As a result, women feel less optimistic about their futures and more like outsiders in agricultural industries that cannot operate without women's labour. While women have increased their efforts in the wake of climate and policy changes, state and industry bodies continue to operate in an environment that ignores gender. The continuing male dominance of industry bodies, the lack of gender mainstreaming in climate and agricultural policies, and the total disregard of the need for supportive social policies for families and communities making significant adjustments confirms that gender equality is unimportant to those with the power to effect change. As we argue in this chapter, women's work on farms has become more essential at the same time as it has become more hidden and the additional factor of a changing climate is significantly increasing the uncertainty experienced by farm families. Attention to gendered relations is essential if we are not only to avoid the continuation of the invisible farmer tradition, but also to ensure food security. Feminist scholars must continue to expose gender inequalities in agriculture and food production in the developed world and argue strongly for gender mainstreaming in agricultural and climate-related policies.

**Acknowledgements:** We wish to acknowledge the Australian Research Council Linkage Program (LP130100676) and the Geoffrey Gardiner Foundation who funded the research on which this was based. We also acknowledge the Regional Wellbeing Survey group, University of Canberra who allowed us to partner on their annual national survey in order to access data for the area and industry that formed the basis of our research. We thank Neil Diamond for assistance with statistical analysis.

# References

Alston, M. (2000). *Breaking through the grass ceiling: Women, power and leadership in rural Australia.* Harwood Publishers.

Alston, M. (2012). Rural male suicide in Australia. *Social Science and Medicine, 74*(4), 515–522. https://doi.org/10.1016/j.socscimed.2010.04.036

Alston, M. (2015). *Women and climate change in Bangladesh.* Routledge.

Alston, M., & Akhter, B. (2016). Gender and food security in Bangladesh: The impact of climate change. *Gender, Place and Culture, 23*(10), 1450–1464. https://doi.org/10.1080/0966369X.2016.1204997

Alston, M., & Whittenbury, K. (2014). Social impacts of reduced water availability in Australia's Murray Darling Basin: Adaptation or maladaptation. *International Journal of Water*, 8(1), 34–47.

Alston, M., Whittenbury, K., Haynes, A., & Godden, N. (2015). Are climate challenges reinforcing child and forced marriage and dowry as adaptation strategies in the context of Bangladesh? *Women's Studies International Forum*, 47, 137–144. https://doi.org/10.1016/j.wsif.2014.08.005

Alston, M., Whittenbury, K., Western, D., & Gosling, A. (2016). Water policy, trust and governance in the Murray-Darling Basin. *Australian Geographer*, 47, 49–64. https://doi.org/10.1080/00049182.2015.1091056

Alston, M., Clarke, J., & Whittenbury, K. (2017). Gender relations, livelihood strategies, water policies and structural adjustment in the Australian dairy industry. *Sociologia Ruralis*, 57, 752–768. https://doi.org/10.1111/soru.12164

Alston, M., Whittenbury, K., & Clarke, J. (2018). Limits to adaptation: Reducing irrigation water in the Murray-Darling Basin dairy communities. *Journal of Rural Studies*, 58, 93–102. https://doi.org/10.1016/j.jrurstud.2017.12.026

Australian Broadcasting Commission Rural. (2016, August 17). *Sons still dominate family farming inheritance*. http://www.abc.net.au/news/rural/2016-08-17/family-farming-inheritance/7756246

Bock, B., & Shortall, S. (Eds.). (2017). Gender and rural globalization: An introduction to global perspectives on gender and rural development. In *Gender and rural globalization: International perspectives on gender and rural development* (pp. 1–7). CABIP.

Brandth, B. (2002a). Gender identity in European family farming: A literature review. *Sociologia Ruralis*, 42(3), 181–200. https://doi.org/10.1111/1467-9523.00210

Brandth, B. (2002b). On the relationship between feminism and farm women. *Agriculture and Human Values*, 19, 107–117. https://doi.org/10.1023/A:1016011527245

Brandth, B., & Haugen, M. S. (1998). Breaking into a masculine discourse. Women and farm forestry. *Sociologia Ruralis*, 38(3), 427–442. https://doi.org/10.1111/1467-9523.00087

Bryant, C. R., Delusca, K., & Sarr, M. A. (Eds.). (2016). Introduction. In *Agricultural adaptation to climate change* (pp. 1–10). Springer.

Dempsey, K. (1992). *A man's town: Inequality between women and men in Rural Australia*. Oxford University Press.

Fletcher, A. J., & Kubik, W. (Eds.). (2016). Introduction: context and commonality: Women in agriculture worldwide. In *Women and agriculture worldwide: Key issues and practical approaches* (pp. 1–10). Routledge.

Food and Agriculture Organization. (2010). *Roles of women in agriculture*. The SOFA Team and Cheryl Doss.

Food and Agriculture Organization. (2013). *The female face of farming*. Food and Agriculture Organization of the UN. http://www.fao.org/gender/infographic/en/

Haugen, M. S., & Brandth, B. (2017). Gender identities and divorce among farmers in Norway. In B. Bock & S. Shortall (Eds.), *Gender and rural globalization: International perspectives on gender and rural development* (pp. 185–197). CABI International.

Hussey, K. (2014). Using markets to achieve environmental ends: Reconciling social equity issues in contemporary water policy in Australia. In T. Fitzpatrick (Ed.), *International handbook on social policy and the environment*. Edward Elgar Publishing.

Intergovernmental Panel on Climate Change. (2014a). *Climate change synthesis report: Summary for policymakers*. https://www.ipcc.ch/site/assets/uploads/2018/02/AR5_SYR_FINAL_SPM.pdf

Intergovernmental Panel on Climate Change. (2014b). *Climate change: Impacts, adaptation and vulnerability.* http://www.ipcc.ch/pdf/assessment-report/ar5/wg2/ar5 _wgII_spm_en.pdf

International Fund for Agricultural Development. (2015). *Food security: A conceptual framework.* http://www.ifad.org/hfs/thematic/rural/rural_2.htm

Knuttila, M. (2016). *Paying for masculinity: Boys, men and the patriarchal dividend.* Fernwood Publishing.

Murray-Darling Basin Authority. (2010). *The Basin plan.* http://www.mdba.gov.au/ basin_plan

Murray-Darling Basin Authority. (2012). *The Murray Darling Basin plan.* Commonwealth of Australia. https://www.legislation.gov.au/Details/F2012L02240

National Farmers' Federation. (2012). *Farm facts.* http://www.nff.org.au/farm-facts.html

O'Hara, P. (1998). *Partners in production? Women, farm and family in Ireland.* Berghahn Books.

Pattnaik, I., Lahiri-Dutt, K., Lockie, S., & Pritchard, B. (2017). The feminization of agriculture or the feminization of agrarian distress? Tracking the trajectory of women in agriculture in India. *Journal of the Asia Pacific Economy, 23*(1), 1–18. https://doi.org /10.1080/13547860.2017.1394569

Pinstrup-Andersen, P. (2009). Food security: Definition and measurement. *Food Security, 1*, 5–7.

Preibisch, K. L., & Grez, E. E. (2010). The other side of el Otro Lado: Mexican migrant women and labor flexibility in Canadian Agriculture. *Signs, 35*(2), 289–316. https://doi .org/10.1086/605483

Price, L., & Evans, N. (2006). From 'as good as gold' to 'gold diggers': Farming women and the survival of British family farming. *Sociologia Ruralis, 46*(4), 280–298. https:// doi.org/10.1111/j.1467-9523.2006.00418.x

Prugl, E. (2010). Feminism and the post-modern state: Gender mainstreaming in European Rural Development. *Signs, 35*(2), 447–476. https://doi.org/10.1086/605484

Riley, M. (2009). Bringing the 'invisible farmer' into sharper focus: Gender relations and agricultural practices in the Peak District (UK). *Gender, Place & Culture, 16*, 665–682.

Rosenfeld, R. (1986). US farm women. *Work and Occupations, 13*(2), 179–202. https://doi .org/10.1177/0730888486013002001

Sachs, C. (1983). *The invisible farmers: Women in agricultural production.* Rowman & Allanheld.

Sachs, C., & Alston, M. (2010). Global shifts, sedimentations and imaginaries: An introduction to the special edition of women and agriculture. *Signs, 35*(2), 277–288. https://doi.org/10.1086/605618

Sheridan, A., & Haslam-McKenzie, F. (2009). *Revisiting missed opportunities: Growing women's contribution to agriculture* (Pub. No. 09/083). Rural Industries Research and Development. http://awia.org.au/wp-content/uploads/2015/05/Revisiting-missed -opportunities.pdf

Shortall, S. (1999). *Women and farming: Power and property.* Palgrave Macmillan.

Shortall, S. (2006). Gender and farming: An overview. In B. B. Bock & S. Shortall (Eds.), *Rural gender relations: Issues and case studies* (pp. 19–26). CABI Publishing.

Shortall, S. (2017). Rurality and gender identity. In B. B. Bock & S. Shortall (Eds.), *Gender and rural globalization: International perspectives on gender and rural development* (pp. 162–169). CABI International.

Shortall, S., & Alston, M. (2016). To rural proof or not to rural proof: A comparative analysis. *Politics and Policy, 44*, 35–54.

Timbal, B., Abbs, D., Bhend, J., Chiew, F., Church, J., Ekström, M., Kirono, D., Lenton, A., Lucas, C., McInnes, K., Moise, A., Monselesan, D., Mpelasoka, F., Webb, L., & Whetton, P. (2015). *Murray Basin cluster report, climate change in Australia Projections for Australia's natural resource management regions: Cluster reports* (M. Ekström, P. Whetton, C. Gerbing, M. Grose, L. Webb, & J. Risbey, Eds.). CSIRO and Bureau of Meteorology.

To, H. (2017). *Review of water reform in the Murray-Darling Basin. Appendix 3: Socio-economic changes in the Basin.* Wentworth Group of Concerned Scientists. http://wentworthgroup.org/wp-content/uploads/2017/11/Wentworth-Group-Review-of-water-reform-in-MDB-Nov-2017.pdf

Torquebiau, E., Tissier, J., & Grosclaude, J-Y. (2015). How climate change reshuffles the cards for agriculture. In E. Torquebiau (Eds.), *Climate change and agriculture worldwide* (pp. 1–16). Springer.

Whatmore, S. (1991). *Farming women: Gender, work and family enterprise.* Macmillan.

Whetton, P. (2017). *Review of water reform in the Murray-Darling Basin. Appendix 4: Climate change in the Murray-Darling Basin.* Wentworth Group of Concerned Scientists. http://wentworthgroup.org/2017/11/review-of-water-reform-in-the-murray-darling-basin/2017/

Williams, J-A. (1992). *The invisible farmer: A report on Australian farm women.* Commonwealth Department of Primary Industries and Energy.

World Bank. (2012). *World development report 2012: Gender and development.* The International Bank for Reconstruction and Development/The World Bank. http://siteresources.worldbank.org/INTWDR2012/Resources/7778105-1299699968583/7786210-1315936222006/Complete-Report.pdf

World Bank. (2017). *Help women farmers get to 'equal'.* http://www.worldbank.org/en/topic/agriculture/brief/women-farmers-getting-to-equal

# Reflection on Chapter 7

## What Is Man-Made Can Be Unmade

*Alana Johnson AM*

**Alana Johnson**, *farmer, feminist activist, founding member of Australian Women in Agriculture, Director of the Goulburn Murray Water Corporation, Chair of the Victorian Catchment Management Council and former Director of the national Rural Industries Research and Development Corporation. Alana is widely recognized for her work in rural development and gender equality. In January 2020 she was appointed a Member of the Order of Australia for her services to women. In 2019 Alana received a Distinguished Alumni Award from La Trobe University. She is also a graduate of the Australian Rural Leadership Program and the Australian Institute of Company Directors. Alana has been a member of the Women's Advisory Councils to the Federal and Victorian state Ministers for Agriculture. She was the Australian Rural Women's Award runner up for 2010 and was named in the inaugural 100 Women of Influence in Australia by the Australian Financial Review.*

### Introduction

The chapter by Alston, Clarke, and Whittenbury tracks the story of my life. Having been raised in a multi-generational farming family in the southeast of Australia in the 1960s, I was the product of a conservative Catholic gender prescription, production focused farming practices with virtually no consideration of environmental impacts, and an agricultural industry that rendered the participation and contribution of women invisible.

In 1974, the Whitlam-led Labour government made university education free in Australia, which resulted in a flood of young rural women like me to the major cities during the peak of the second wave women's liberation movement. While many young rural men stayed home to work on the farm or follow construction or technical careers, the university-educated young women arrived back in rural Australia in the early 1980s with newfound knowledge, ideas, behaviours, and

DOI: 10.4324/9781003089209-15

expectations. This provided the springboard for what was to become the Rural Women's movement and also the Landcare movement in Australia.

The past four decades my life's work have seen slow yet persistent progress toward women in agriculture gaining a voice, public and policy recognition of women's work and participation in the sector, increasing numbers of women having a seat at the decision-making tables, social and environmental factors being included in a quadruple bottom line for production agriculture, and farmers leading climate action in our nation of prolonged government climate change denial.

My work has been the work of many women who have stepped up into leadership positions and who, in 1993, established Australian Women in Agriculture as a peak national body and in 1994, organized the First International Conference for Women in Agriculture, hosted in Melbourne, Australia.

## Representation

Alston, Clarke, and Whittenbury give an accurate account of the longstanding invisibility of women in agriculture, but finally this is changing in the agriculture and environment sectors in Australia. The 2014 appointment, in my state of Victoria, of the first female Minister for Agriculture has since been followed by the appointment of two subsequent female Ministers for Agriculture. This has resulted in a significant claim of women in this space and a challenge to the dominant male construct. Together with the female Minister for the Environment, these Ministers initiated legislation to require all state government instrumentalities, such as the Water Corporations and Ministerial Advisory Councils, to have 50% of board positions filled by women. This has created long-awaited fundamental changes for rural women.

In 2018, Fiona Simson was the first woman to be appointed Chair of the National Farmers Federation—the nation's peak farmer representative body. In 2020, Emma Germano was elected as President of the Victorian Farmers Federation, the farmers' representative body in my state. Both of these women have high media profile and are greatly respected by governments and the farming sector.

The Invisible Farmer project—backed by Melbourne Museums, the Australian Broadcasting Corporation, and Melbourne and Monash Universities—began in 2017 as the largest-ever study of Australian women on the land. The project captured the nation creating new histories of rural Australia and revealing the hidden stories of women and farming. Directed by the well-known adage "you can't be what you can't see," the Invisible Farmer project has rendered visible women as farmers and agri-professionals and inspired a new generation of women in agriculture.

## Who Are the Farmers?

As cited in this chapter, some decades ago, Sally Shortall wrote that often women's work supports men's career choice to be farmers and that many farms would not survive without the labour and income of women. This still plays out on many

Australian farms as men "hang on" to unviable dairy farms or grazing enterprises, often on marginal land. Women, through family circumstances, get locked into a desperate precarious cycle in which providing for their family and supporting the farm (and their farmer husband) is their "lot."

This has been particularly evident during the increasing periods of drought when farmers are portrayed to the public as "doing it tough" and as not having the finances to feed their starving stock or their families. Media funding appeals realize both public sentiment and money for farmers that too often don't have management or contingency plans for periods of drought and associated high irrigation water prices. These "poor farmers" are seen as victims to the vagaries of the climate.

This continues to be a reality and the public view, despite the many farmers who are proactively planning for the changing climate, who are taking actions to reduce carbon emissions, who are pursuing environmental sustainability, and who are adapting their business model to be productive and viable in the changing climate.

## Gender Privilege

There is a changing perception of who is a farmer, and we see far more images and reporting of women doing farm work and business today, but there is still a hidden privileging of men's farm work within the domestic realm. While women will organize their farm work around childcare, household duties, food buying and preparation, and children's school related activities, men will still often do farm work from dawn to dusk and come home at the end of the day expecting to be fed. Australian women on average perform more domestic work than their male partners and this is often more exaggerated in farming families. Men's farm work is perceived as the most important work. Many farming men will work 12-hour days not necessarily because it needs to be done, but just because there is always work to be done. It is the core of their identity and how they operate in the world, and most often their wife/partner just complies—albeit with a range of complaints to their children and other women.

Thankfully we can see a shift in this current generation with both young women and men overtly engaged in negotiating their domestic, financial, and relationship arrangements based on the expectation of more equality and autonomy for both parties. These young women have expectations of men's contribution and role, and are not backward in expressing this.

## Who Owns the Farm?

Patrilineal inheritance has been normal practice in farming families in Australia and the resulting gender equity issues are well known. Rightly, there is increasing equal sibling inheritance of farming land, but this is not necessarily a positive solution as it can create complex issues for farm management. Often the financial arrangements to pay non-farming siblings an annual lease or capital sum means the non-farm income of a woman married to a male farmer is utilised to offset the financial distribution made to his siblings.

When the management of the farm property and business passes to the next generation, widowed farm women can be required by the family to leave the farm homestead, and their finances can be in the hands of their adult children. This does not always play out well in terms of older women's autonomy or future living circumstances.

On the bright side, more young people farming in Australia are leasing land for their farming enterprise and this includes many young, well-educated women farmers. There are now other viable pathways for women to enter farming other than through marriage.

It needs to be noted that the continual decrease of smaller family farms over past decades has been coupled with an increase in corporate owned farms. This is particularly evident in irrigation as large horticulture and dairy businesses have refined business strategies to accommodate fluctuations in productivity and profit. It can be argued that the move to corporate farming further excludes women in agriculture as the irrigation and farm managers, as well as the company directors, are overwhelmingly men.

## The Dairy Industry in the Goulburn Murray Irrigation District

As noted by Alston, Clarke, and Whittenbury, the effects of climate change are pronounced in the irrigated dairy sector.

Irrigation water was privatized in 2007 and the price of water through the water markets is dictated by water availability and seasonal variations. Due to climate change, over the past 20 years there has been a 40 percent decline in water inflows into the Murray River catchment and this decline will continue. The resulting precariousness of water availability and cost of water has meant the number of dairy farms in the Goulburn Murray Irrigation District (GMID) has reduced by 50 percent in the past 20 years. The implications for the future of irrigated agriculture are stark.

Fortuitously, in 2021, GMID dairy farmers had one of the best years for decades. Water is readily available and affordable, and milk prices are high. It may seem like a "boom" time—but not for all. Due to the COVID pandemic there have been severe labour shortages and consequently a greater labour burden on dairy farming women.

The dairy sector has historically been the leader in acknowledging women's contribution to agricultural industries. It is accepted that dairy farms rely on husband/wife partnerships and that women are a core part of the business. Consequently, women have long been part of the industry events, the industry award systems, and increasingly on representative bodies. Yet although women are recognized in the sector, their labour is often taken for granted as part of the farm domestic arrangements.

## An Impact of Climate Change

The increasing impacts of climate change have seen an escalation of challenging times and crisis in the farming sector. Droughts, fires, storms, and floods as well as the unavailability of affordable water have taken a toll on the physical and mental health of farmers. Being a "farmer" is bound up in the identity of most farming men, so it is not surprising the mental health of farm men and suicide risk are of great concern as the financial and work stress on farms escalates. The result is women increasingly "holding the family together"—acting to protect their partner's mental health, relieving the workload pressure, especially the administrative work, and in some cases experiencing and excusing heightened volatility and even violence and abuse by their partner.

In an era where increased agency for women is the norm across society, the impacts of climate change on the farming sector are often resulting in the opposite for farm women— increased work burden and even increased fear for their safety.

## Women's Activism

Issues important to women—whether they be the appalling inadequacy and unaffordability of childcare for rural families, the lack of rural family violence services, the cost of irrigation water, women's safety in rural workplaces, or achieving carbon-neutral farming—still do not have the same platform as "man speak." In the 1990s, Australia led the world in rural feminist action with the formation of the peak advocacy body, Australian Women in Agriculture (AWiA), and government-funded programmes for women to be recognized, to be heard, to be seated at the decision-making tables, and to have a say in policy development. Over the decades there has been some shift in perception with more women visible in the sector and at board tables (although far from 50/50). But women on farms are still mostly invisible and unheard. The local irrigation representative committees for the Goulburn Murray Water Corporation have less than 10 percent women although women are actively encouraged to apply.

Traditional gender relations are still at play as men take on the visible roles and women take care of families and support the men.

The peak body, AWiA, and the government funded Rural Women's Coalition are deafeningly silent with regards to gender relations and gender analysis. Calling oneself a feminist and taking overt action for gender equality is still seen by many men and women as extreme and anti-men (with whom the women are often in domestic and work relationships). Farm women's identity is still as bound up in the traditional gender construct as is that of male farmers. Women in the sector seem to think that hard work and proving yourself as a capable farmer will somehow lead to the structural changes that need to take place for gender equity in the agriculture sector.

## Women's Agency

Continued "passivity" by women in agriculture is accompanied by continued research and reporting of women in agriculture as victims of the "system," of the patriarchy, and now of the impacts of climate change.

Women who have achieved profile and positions of leadership and decision making in the agriculture sector have learnt to adapt and even comply with the patriarchal institutions in which they have been "allowed" to participate. Like most female politicians, these women leaders avoid the repercussions of "playing the gender card" because of the backlash and the impact it will have on their careers. Consequently, the overt pursuit of gender equality, which is interpreted as challenging the "man made" processes and structures in the agriculture sector, is not part of the national narrative.

In Australia over the past decade, we have witnessed an accelerating shift in power. The exposures within national institutions such as child abuse and cover-ups in the Catholic church, illegal activities in the banking system, abuse and neglect in the aged care system, the funding routes by the party in government and obscure political donations has resulted in a public loss of trust and an increase in ordinary people becoming activist in political and government processes. People no longer believe that "G"overnment will make things better or act for the common good and they are increasingly taking collective action.

The year 2021 saw a huge nationwide rising of women with the Women's March 4 Justice following exposures of the abuse of women and girls in schools, the sports sector, the arts sector, and within Federal Parliament House. The time is nigh for women in agriculture to also step into their power, to no longer accept being victims of the gender discrimination in the sector and to challenge farm gender arrangements. Climate change is exacerbating gender inequity for women on farms but change for women will not occur unless demanded by women.

## Conclusion

The chapter by Alston, Clarke, and Whittenbury correctly reports that "complex gender negotiations occurring in the private family sphere [are] being ignored by industry and government" and that "state and industry bodies continue to operate in an environment that ignores gender." They highlight that the lack of supportive social policies confirms that gender equality is "unimportant to those with the power to effect change." Thus, it has always been so!

While from a social justice perspective, gender equality should be a priority, those with "the power to effect change" have never taken the lead. The change required will not happen until women in agriculture step into their power, and on behalf of all women and girls become demanding, disrupt the patriarchal practices, and be vocal on farms, around the board tables, and in the corridors of parliament. Gender compliance and waiting for men to bring about change will not achieve results. What is man-made can be unmade and remade, but this will have to be led by women themselves.

# Conclusion

## Welcoming a New Climate Future

*Maureen G. Reed and Amber J. Fletcher*

The message is clear. Even in rich countries of the Global North, climate change has different effects across *and within* communities. Furthermore, people continue to be left out of the planning and policy processes that can help them mitigate and adapt to a changing climate. These processes are not restricted to climate change policies, since changing climate conditions demand that we reconsider how (and how much) we extract natural resources; how and with whom we build our social infrastructure and networks; how we interpret and calculate social, cultural, and economic gains and losses; as well as how we build our physical infrastructure and response strategies to mitigate direct and indirect losses when hazard events happen.

By highlighting intersections among gender and other social identity factors, and between identity factors and social structures and power relations, the contributions in this book highlight the diversity of people who are affected by climate hazards and long-term climate change—from pastoralists in Spain to refugees in Australia. Contributors also make visible a range of social threats rarely discussed among academics and policy makers. These include threats to bodily safety, emotional wellbeing, collective agency, and to the places and spaces that matter to people. The authors also illustrate a compelling need to think about how the experiences and knowledges of marginalized people can become more prominent in policy processes that govern natural resource extraction and sustainability, and in practices that regulate activities such as impact assessment and infrastructure development. They argue that more inclusive engagement processes will challenge the status quo of climate impacts and adaptation and build more robust, fair responses to a growing climate emergency.

But the contributions do not rest there. Commentators in this volume draw attention to the fact that climate change is a profoundly human issue. In particular, people "on the front lines" speak eloquently to our humanity when describing the effects of climate change, whether these be felt at an emergency response centre in Canada during a wildfire, in the pasturelands of northern Spain, atop a mountain range in the United States, or at a refugee centre in Australia. These authors remind us that we cannot lose sight of how climate change touches on our everyday lives, our social fabric, and our compassion and need to care for one another.

DOI: 10.4324/9781003089209-16

## Key Insights

Several overarching insights can be derived from the contributions to this volume. A key one is the importance of learning from everyday experience. This insight permeates longstanding feminist research and is no less significant when trying to understand how climate change shapes local experience and action. The operation of gender and intersectional power relationships is subtle—a fact that has shaped ongoing denial of how inequality affects the daily life of marginalized people irrespective of climate change. For example, through their action-research approach in Canada, Rachel Reimer and Christine Eriksen reveal both the subtle and not-so-subtle operation of sexism in mountain professions, despite a pretext of acceptance and inclusion. Alison Criscitiello grounds these insights by describing her experiences of being underestimated or disregarded as a top-performing mountain ranger and guide. These authors link everyday experiences to interpretations of risk and competency, including risks and competencies associated with climate change. The result is that men, particularly White men, by virtue of their gender and racialized position, are considered more likely to be risk-takers and thereby more competent to address the immediate risks associated with climate-related disasters such as avalanches and wildfires.

Together, these authors ask: "Can we imagine—or perhaps reclaim—these environments as a non-gendered space that fosters the emergence of tolerant and inclusive professional mountain and wildland cultures?" From their research, embodied feminist leadership in mountain environments may involve critically interrogating the status quo (hooks, 2000), or it may take the form of the quiet internal cultivation of self-worth and self-confidence. Their question and the insights they derive from addressing it, however, transcend mountain culture with relevance wherever women and gender-diverse people take up leadership roles to address climate change.

This theme of gendered leadership is further reflected in Heidi Walker's intersectional analysis of a major wildfire event. Public debriefing events in the aftermath privileged perspectives from male-dominated sectors, like external fire-fighting agencies, while ignoring the important contributions made by women, Indigenous people, and other local community members. Some community members responded by questioning or challenging this narrow approach to post-disaster learning, and by asserting the relevance of their skills and knowledge for future wildfire preparedness.

Similarly, Federica Ravera, Elisa Oteros-Rozas, and María Fernández-Giménez use a life history approach to document the everyday experiences of women pastoralists in Spain, explaining how women pastoralists expressed their understanding and experience of climate hazards through their bodies, their time, and their emotions. Together with pastoralist Lucía Cobos's poignant and poetic reflection on the emotive and bodily experience of climate change, we are reminded that climate change has very material, felt effects. Climate change is visceral and real, especially for those most closely connected to the land. As Cobos says: "the quality of my sleep and the intensity of my sleepless nights depend on the heat and rain." Agency is found here, too: The authors describe how women pastoralists have mitigated and adapted to climate change through actions that reflect both traditional conservation and innovation.

The actions of everyday life, however, are not independent of the social structures and power relations within which people are embedded. An ongoing criticism of the application of intersectionality has been its focus on individual identities to the exclusion of systemic power imbalances. Insights are gained from authors who explicitly addressed this gap. For example, Heidi Walker's chapter demonstrates how to apply the concept of intersectionality and undertake a nuanced, multi-level analysis. Her contribution provides a practical strategy for analyzing climate hazards, which reveals how social identity attributes such as gender, race, socio-economic status, and family status intersect with one another as well as with powerful discourses and social structures to shape the experiences of wildfire for residents in a northern Canadian community. Her efforts not only advance our understanding of feminist theory in relation to climate hazards, but also suggest practical recommendations for building inclusive and meaningful processes that can empower communities to plan their own strategies for mitigation and adaptation.

This multi-level approach is also adopted by Shefali Juneja Lakhina and Christine Eriksen, who explicitly consider *whose* narratives, experiences, and practices count as Australia reckons with the extreme cascading impacts of the climate crisis. More specifically, they draw attention to the adaptation strategies of a social group frequently overlooked—newly arrived and recently settled families, women-headed households, and individuals from refugee and migrant backgrounds. Their research uncovered the systematic exclusion of newcomer families to the Illawarra region of Australia, despite their lived experience and valuable expertise on climate change adaptation and food security. Furthermore, the authors find that these systemic issues are pervasive in workplaces and the wider community, and are replicated across agencies and authorities at multiple levels. Their findings are reinforced by Sherryl Reddy, who writes from her extensive experience in the humanitarian and emergency services sector. She tells of deeply rooted disparities in how those in positions of power determine whose knowledge is taken seriously when decisions are being made.

Multi-level analyses in this volume have revealed another important insight. Formal action on climate change has yet to challenge masculinist structures and power relations associated with resource extraction and impact assessment. In Canada, regulatory change on environmental issues—such as that documented by Leah Levac, Jane Stinson, Deborah Stienstra, and discussed further by Anna Johnston—continues to favour longstanding power relations that too frequently exclude meaningful engagement of affected communities, and key groups within those communities. The development and application of a community vitality index identified important aspects of wellbeing that extend far beyond typical measures. Levac, Stinson, and Stienstra's methodology illustrated how community members connect environmental health with culture and wellbeing. Their chapter illustrates "the valuable contributions women with diverse identities can make when they are positioned and recognized as experts." These authors conclude that changes to the *Impact Assessment Act* to include "sex, gender, and other intersecting factors" are hopeful signs that greater inclusion is possible.

Margaret Alston, Josephine Clarke, and Kerri Whittenbury also see the potential for policy to create meaningful change. Their observations about gender relations in Australian agriculture are reinforced by agricultural leader and farmer Alana Johnson, who calls strongly for women's activism and agency to change the status quo of agriculture—particularly as drought stress further extends women's already-stretched labour capacity. Alston and colleagues discuss the impacts of drought on both farm men and women. For men, personal identity is intertwined with the farm, causing mental health problems during times of climate stress. Women, in contrast, become increasingly responsible for providing both emotional and labour support. The authors call for gender mainstreaming to highlight and address the gendered dimensions of water management policy.

Despite the possibilities of policy, Anna Johnston reminds us that wiggle words and Ministerial discretion in Canadian environmental policy will continue to require vigilance and on-going (if exhausting) efforts if communities are to make a dent in the seemingly relentless drive of resource extraction in Canada. The question of inevitability in extraction industries is also addressed by Mary Boyden. Through her extensive experience in the mining sector, Boyden speaks to the importance of impact assessment for Indigenous communities, noting that communities must be involved in decisions about if, and how, an extraction project should proceed. Groups like Keepers of the Circle illustrate the power of Indigenous women to change the way resource extraction is done in Indigenous communities around the world. Drawing on the knowledge of her teachers and guiding Elders, Mary shares the Anishinaabe wisdom of the coming Eighth Fire—a time when humanity will need to choose its way forward.

Moving from mining to the oil and gas sector, Angeline Letourneau and Debra Davidson also suggest a new direction for resource industries. Their chapter demonstrates the deeply engrained norm of "petro-masculinity" in Alberta that currently steers the political discourse of the province towards a precarious climate future. Petro-masculinity is an identity that celebrates "the industrial, self-sacrificing working man" who serves as a male breadwinner, extracting raw resources from the earth. This identity has moved beyond the boundaries of industrial facilities to shape a government imperative toward an economic future founded on petroleum, despite science showing that such a future is unsustainable and likely catastrophic in light of ongoing climate extremes and the global transition away from fossil fuels. Just like Mary Boyden, however, Letourneau and Davidson offer hope. They suggest that groups like 350.org and MenEngage can help marry the dual need for climate action with a reimagined masculinity based on ecological values and empathy.

## Deepening our Understanding

The reflections from people who have direct experience of climate hazards and climate action have deepened our understanding of, and appreciation for, climate change "at the intersections." The deep experiences of women working as mountain guides, pastoralists, farmers, miners, support workers for newcomers, lawyers, and first responders provide us with insights that we cannot glean from government reports, newspaper clippings, or academic theories. These contributions help us not only to "see" but also to hear, feel, and gain empathy for the challenges and actions of those at the front lines of climate change. Their contributions demonstrate that experiential, practice-based, Indigenous, and felt knowledges, among others, complement academic knowledge to better understand the complex social dynamics that characterize our responses and adaptation to a changing climate.

Mountain guide and glaciologist Alison Criscitiello, for example, explains how risk is an embodied concept; women bring different interpretations of risk due to their physical and emotional connections with the natural world. Drawing on her own experiences, Alison explains the "everyday chore" of having to demonstrate her own physical and mental fitness, explain how she assigns risk, and overcome the dominant assumption that masculinity conveys competence. Her lived experience in cold, risky environments combines with her scientific expertise to create complex, hybrid knowledge crucial to understanding climate change.

From the other side of the world, Sherryl Reddy's reflection draws attention to the differential impacts of climate-related hazards for a longstanding yet underserved community—humanitarian refugee entrants who are resettled across the Global North. Women refugees and migrants in Australia have diverse lived experiences, hold traditional and place-based knowledge of the changing climate, and can contribute significantly to community-centred preparedness and response initiatives in collaboration with emergency service organizations. Nevertheless, Reddy explains that these voices are systematically excluded or ignored. This compels us to think about intersectionality and climate change in a new way, laying bare how social systems that are designed to support newcomers and climate action can remain highly discriminatory and disempowering. Speaking as a legal scholar-activist, Johnston confirms Reddy's conclusions in a different context—that is, the well-meaning intentions of legislative change in Canada's *Impact Assessment Act*. Johnston's intervention reminds us that because underlying, discriminatory norms persist and multiple forms of discretion remain, the potential gains from the new legislation remain just that: Potential. And consequently, these gains remain precarious and empowerment remains uncertain.

Writing from "northern" Canada, Nancy Lafleur explains that fire season has always been part of her calendar, but it is growing less predictable and more threatening to lands, "resources," and cultural practices. Lafleur and Lucía Cobos both share stories that intertwine the experiences of climate change with family. Nancy reminds us of the important difference between filling out forms and caring for family when addressing a climate disaster, while Lucia explains an important choice she had to make between her son and her cows. They describe these lessons through the everyday artefacts of the annual lifecycle—a birthday cake; a first tooth; some blankets; and a fire.

These direct experiences remind us that we, as scholars, activists, and policy makers, must seek out diverse groups of people when documenting and recommending environmental and climate change policy and action. Engaging those who may find themselves at the margins is not merely required to understand their situations, but to bring their knowledge and first-hand experience to bear on strategies we will need to adapt to a changing climate. Significantly, the testimonies of the commentators are a reminder that these concerns are not only academic ones, but rather, are rooted in the beliefs, hopes, dreams, and worries of the human heart. Academic articles, while providing important knowledge, have yet to stir meaningful action. Drawing on our humanity just might.

## Looking Ahead…and Toward New Kinds of Actions

Finding common ground seems difficult right now. As we write, the world is entering its third year of the COVID-19 pandemic. Tensions are mounting. In Canada, a convoy of trucks recently made its way across the country and camped out in the capital to protest vaccine mandates—and the protest includes people who uphold dangerous and hateful perspectives like White supremacy. Anti-science rhetoric plagues public discussion of climate change, too. Although awareness and concern about climate change is increasing throughout the world, many people continue to deny the very existence of the problem (Leiserowitz et al., 2021). How should feminist, intersectional analysis proceed in a time of marked anti-science rhetoric and hate-fuelled activism? Rather than a de-politicization of climate change, some feminist scholars emphasize the need to engage with multiple perspectives—to view a problem from more than one angle for a more fulsome understanding and to take actions that are effective and equitable.

So, who cares about climate justice? Of course, we all do. Sometimes the scope of the task seems overwhelming. Lucía Cobos's observation that adaptation is not merely resignation but may also be a form of resistance is an important one. This theme runs through each of the contributions in this volume. Together, contributors remind us that adaptation requires care, compassion, and cultural sensitivity. These "soft skills" bring hard lessons. For example, Nancy Lafleur's reflection reveals how adherence to strict policy and procedure, so often designed from a colonial standpoint, may stand in the way of what people truly need to recover. Her narrative teaches us that more effective wildfire evacuation is that which considers cultural and spiritual values, responding to a wide range of needs beyond the physical.

Contributors to our understanding of masculinity suggest that consciously choosing to subvert hegemonic masculine norms is an act of defiance, or even deviance. As noted in Chapter 6, perhaps innovation can be found by engaging in positive deviance from the status quo (Pascale et al., 2010). Indeed, Alana Johnson emphasizes the importance of such agency and resistance, calling for women in agriculture to challenge gender conventions and claim their power to make positive change. Many contributors who are seeking recognition for their gender identities and associated competencies are highly motivated to work for change. We have only touched on gender and sexual diversity within this volume. We need to better understand how women, Black, Indigenous, and People of Colour (BIPOC), and LGBTQ2SIA+people defy narrow, constraining norms and innovate as they seek to mitigate, adjust, and thrive in a changing climate.

Finally, developing a caring approach to climate resilience goes beyond situating differential vulnerabilities. Rather, we need to better understand and promote transformative practices of care—for self, community (as noted by Lakhina and Eriksen), and friends we have yet to meet. Feminist scholarship and action that draws attention to the intersectional character of climate hazards requires us to not just ask *who* is affected by climate change impacts, but also to engage with empowering and transformative narratives and strategies that reveal how people cope, resist, and adapt across changing landscapes. This is not a call for a single movement. As Lakhina and Eriksen point out, there is no homogenous, well-organized community of belonging—identities can be "fractured, tangential, and context-specific." Mary Boyden, Alison Criscitiello, Lucía Cobos, Alana Johnson, Anna Johnston, Nancy Lafleur, and Sherryl Reddy face challenges that are unique to their gender identity, culture, and geo-social context, and they respond in their own ways to these challenges. "Community strategies" are forged and redesigned daily. Learning from multiple and diverse perspectives can spur transformation, ideally toward a more welcoming climate future.

## References

hooks, B. (2000). *Feminist theory: From margin to center* (2nd ed.). Pluto Press.

Leiserowitz, A., Maibach, E., Rosenthal, S., Kotcher, J., Carman, J., Neyens, L., Marlon, J., Lacroix, K., & Goldberg, M. (2021). *Climate change in the American mind, September 2021*. Yale Program on Climate Change Communication. https://climatecommunication .yale.edu/publications/climate-change-in-the-american-mind-september-2021/toc/2/

Pascale, R., Sternin, J. & Sternin, M. (2010). *The power of positive deviance: How unlikely innovators solve the world's problems*. Harvard Business Press.

# Index

international humanitarian protection
56–57
intersectionality/intersectional: analysis
4, 16, 18–19, 26–27, 39–40, 47, 49, 66;
application of 4, 66–67, 123, 138, 200;
approach 4–7, 18–20, 132; and critical
community engaged scholarship 66–67;
framework 5, 49; inclusion 60–62;
lens 3, 22, 39, 47, 49, 59, 67, 84–86;
marginalization 8, 16, 18, 73, 77, 93;
power and privilege 6, 56; research 5,
7, 17, 20, 39; social factors 6; structure
factors 4, 6; theory of 7, 20, 39
irrigation water 176, 194–196

Jacobs, F. 26

Kaijser, A. 20, 132, 134
Kemess North Copper-Gold Mine 87
Kiama 43
Kimmel, M. 93
Klein, R. 96
Klocker, N. 41–42
knowledge: community 85–86; cultural 49,
59; diverse 11, 26, 40; ecological 130;
Indigenous 40, 66, 72–73, 77, 79, 85–86;
intergenerational 47; local 6, 77; traditional
40, 46–47, 56, 58, 133, 137, 202
Kronsell, A. 5, 20, 132, 134

Lac La Ronge Indian Band (LLRIB)
17–19, 21, 24–25
Lakhina, S. J. 7–8, 56, 59, 61, 200, 204
landscape 96, 100, 102, 111, 114, 120,
128–129, 134, 152, 170; Australian 38,
49; changing 40, 42–43, 47–49, 204;
extractive 94; rural 42, 94, 124, 137;
unfamiliar 39, 47, 49
La Ronge 16–21, 23–26, 32, 37
Las Cumbres 126, 145, 148
Latosuo, E. 153, 162
leadership 149–167; feminist 10, 151, 163,
170, 199; institutional 59; women's
70–71
LGBTQ2S+ 65, 68, 72–73, 76
LGBTQ2SIA+ 153, 161, 163, 204
Liberia 44, 46
Lomborg, B. 94
Lower Churchill Hydroelectric Generation
Project 87
Lower Churchill Joint Review Panel 69

Mackenzie Gas Project 87
Magnusdottir, Gunnhildur Lily 5

masculinity: definition 91–92; ecological
103; and environment 94; and fossil fuels
95; hegemonic 8, 92–95, 98, 104, 175;
hypermasculinity 90, 97–99, 101–103,
158, 160, 162; literature 91–95; mountain
151; oil and gas sector 8, 100–101, 201;
performative 150, 157–158, 161–163;
petro-masculinity 9, 91, 95, 101–103,
201; rural 150; toxic 91–94, 101
masculinity contest culture (MCC) 152
media critical frame analysis 22
mental health 6, 10, 19–20, 69, 71, 127,
162, 175, 184–187, 196, 201
methodology 16–17, 43, 66–67, 73, 124,
152, 200; inquiry 154–155
migrant 40–42, 65, 137–138, 202;
backgrounds 39, 41, 47, 49, 56, 58–59,
61–62, 200
Miller, G. E. 97
Mista-Shipu 69
Mokami Status of Women Council
(MSWC) 69
Montreal Lake 35
Most Valuable Backcountry Ranger 169
mountain professions: climbing 152–153;
effects of climate change 198–199;
gender relations in 150–151
Multicultural Community Liaison Unit
(MCLU) 48, 59–60
Multicultural Liaison Unit 48
multilevel analysis 9, 16, 20–22, 24, 27,
200, 203
Murray-Darling Basin (MDB) 42, 173,
176–178, 186–187
Murray-Darling Basin Plan (MDBA) 177
Musk, E. 94

NALCOR 69
National Report on Missing and Murdered
Indigenous Women and Girls 115
Natural Resources Canada 96
Neimanis, A. 122, 127
New South Wales 40–41, 43, 58–61
non-government organizations (NGOs)
57–58
Notley, Rachel 99
Nowroozipour, F. S. 42
NSW State Emergency Services (SES)
48, 59
NVivo 17, 178

*Oilweek* magazine 97, *97*
Olympic National Park 169–170
Ontario 111–112, 114

For Product Safety Concerns and Information please contact our EU
representative GPSR@taylorandfrancis.com
Taylor & Francis Verlag GmbH, Kaufingerstraße 24, 80331 München, Germany

www.ingramcontent.com/pod-product-compliance
Lightning Source LLC
Chambersburg PA
CBHW060259220326
41598CB00027B/4167

9 781032 316857